The
CD-I DESIGN
Handbook

GW00362684

The CD-I Series

Introducing CD-I
This book provides a comprehensive non-technical overview of CD-I and is aimed at anyone interested in the implications of this revolutionary technology.

The CD-I Production Handbook
Aimed at the video and audio production teams involved in making CD-I titles, this book contains detailed examples exploring the concepts and central issues involved.

The CD-I Design Handbook
Aimed at designers with a technical background but no knowledge of CD-I, this book provides detailed, practical coverage of CD-I design. The different stages of CD-I design are richly illustrated with a wealth of examples.

The CD-I Programmer's Handbook
Aimed at CD-I programmers, this book is the standard programming manual on CD-I.

The official guide to CD-I Design
from Philips Interactive Media Systems

The CD-I DESIGN *Handbook*

PHILIPS IMS

ADDISON-WESLEY
PUBLISHING
COMPANY

Wokingham, England · Reading, Massachusetts · Menlo Park, California · New York
Don Mills, Ontario · Amsterdam · Bonn · Sydney · Singapore
Tokyo · Madrid · San Juan · Milan · Paris · Mexico City · Seoul · Taipei

© 1992 Philips Electronics UK Ltd

Many of the designations used by manufacturers and sellers to distinguish
their products are claimed as trademarks. Addison-Wesley has made every
attempt to supply trademark information about manufacturers and their
products mentioned in this book. A list of the trademark designations and
their owners appears on p. x.

The photographs on the following pages are produced with permission and
© Hulton-Deutsch Collection: 7 (Swan Lake), 8 (Richard Burton), 9 (Frankie,
the lion cub), 11 (The Deposition) and 12 (Michael Chang).
Cover designed by Chris Eley
and printed by The Riverside Printing Co. (Reading) Ltd.
Text designed by Valerie O'Donnell
Typeset by Electronic Type & Design, Oxford in 10/13 Candida and Futura
Printed in Great Britain by William Clowes, Beccles, Suffolk

First printed 1992.

ISBN 0–201–62749–3

Philips and Addison-Wesley would like to acknowledge TMS,
Computer Authors for the initial draft, and Carol Atack for carrying
out the final editing of this book. We would also like to thank the
many reviewers who have contributed valuable comments as the
project progressed and especially Clive Shepherd of EPIC Interactive
and David Matthewson.

British Library Cataloguing in Publication Data
A catalogue record for this book is available from the British Library.

Library of Congress Cataloging in Publication Data

Preface

This book describes the CD-I (Compact Disc-Interactive) design process from initial concept to the beginning of production, covering:

- working up a concept
- visualizing the structure and flow of material on the disc
- storyboarding and scripting the idea
- balancing the use of resources
- identifying and estimating production resources

Both the creative possibilities and the market potential of CD-I are discussed, with reference to existing programmes.

CD-I is a new technology, and the very concept of interactive media is new to many of those wishing to design and produce programmes that exploit its many possibilities. This book contains an introduction to CD-I technology aimed at designers whose need is to understand what they can do with the technology rather than the most intricate details of how it works. It also puts CD-I into context as one of the most exciting interactive media to have been developed over recent years.

Although the book attempts to keep technical jargon to a minimum, it is important that those who plan to work with CD-I are familiar with the specialist vocabulary used to talk about it. Unusual terminology is defined both within the text and in the glossary at the end of this book. The sections on CD-I technology cover:

- the player
- the kinds of data that can be used, including sound, images and text
- getting data on to the disc in the most effective manner
- the output of data from the player
- input devices

However, the hardware is only the beginning: designers also need to consider the way their programmes will look to the viewer, and the way the viewer will interact with the program. Chapter 3 looks at this, summarizes the findings of research on the interaction between people and the machines they use, and offers advice to would-be CD-I designers.

Modelling the design process

CD-I is attracting the interest of producers and developers from a wide range of industries, with differing skills and backgrounds. Designing an effective CD-I programme requires many of these skills and so a team working on a CD-I title is likely to consist of a range of specialists unusual in older forms of media. Chapters 8–10 consider how to put together a CD-I design team and how best to manage the design process.

However, because CD-I design is a new process, any model of it can be based on only a few real-life examples. Experience gained from such examples is an important part of this section of the book. Although detailed case studies are given, future CD-I producers may find that the scale of their project is such that the model created in this book provides general assistance rather than an exact blueprint for their efforts. The conclusions drawn aim to reflect the different scales of CD-I development, which can all lead to the eventual production of exciting CD-I programmes.

As the technology continues to develop, with new authoring tools and editing facilities becoming available, different methods of working may become more appropriate. The multimedia nature of CD-I (see page 2) means that no one industry can provide a completely appropriate model for designers of CD-I programmes. However, this book uses the video/television production industry as a model for most of the process, with a nod to the computer entertainment industry for the parts of the process that are most technology-bound.

Who needs to know about CD-I design?

The production of CD-I titles requires the cooperation of team members with many different skills. Although the various tasks and disciplines involved in the process may appear to be quite separate, it is important that everyone has a full understanding of what other people are doing. Each participant, from producer to programmer, needs to relate his or her task to those of the others, which are in turn subordinated to the ultimate goal of making the interactive compact disc. For this reason, this book is intended to be read not only by people who are actively designing CD-I discs, but also by those who are engaged in other ways in the design, development and production process.

People directly involved in CD-I design, whether as writers, graphic designers, video and audio technicians, editors or managers can discover from this book the limits and possibilities of the medium they are designing for, and see how their individual role fits into the overall picture of CD-I design.

Producers and assistants can gain a necessary understanding of the sorts of process they are taking on. The book will tell them how CD-I development differs from that of other media, and should give valuable insights into the effective management of the creative and design process.

Project managers should likewise benefit from a greater understanding of the CD-I design process, and of the technology itself.

Without this knowledge the management task would become impossible, especially when difficult decisions need to be made about allocating space on the disc for various types of media.

Software managers and programmers can read this book for an insight into the creative aspects of the design that they will be developing and implementing.

CD-I developers will also find the reference work *Compact Disc Interactive Media Full Functional Specification* invaluable. This authoritative volume, referred to in the present work as the Green Book, contains the final word on all aspects of CD-I technology. It is available only with full CD-I development systems.

Using this book

The present book is intended to be a practical guide to CD-I design, and you can and should dip into it for information on specific topics. Those new to CD-I and interactive media will find the early chapters particularly useful.

Chapter 1 The potential of CD-I
Explains what CD-I is, considers who is and who should be using CD-I as a medium for their products, and looks at where CD-I fits in with other multimedia technologies.

Chapter 2 CD-I as a medium
Considers how interactive media differ from linear ones and the design implications, where CD-I will be used and what market opportunities are available, the different genres of CD-I titles, and ways of developing an original CD-I concept.

Chapter 3 Designing for interaction
Looks at the interactive nature of CD-I and the implications this has for programme design and design methodology.

Chapter 4 The interactive interface
Considers the importance of the interface in facilitating communication between users and the programme, and the devices that form part of the interface.

Chapter 5 Using video with CD-I
Provides an overview of image formats for CD-I and their impact on CD-I design.

Chapter 6 Using audio with CD-I
Provides an overview of ways in which the CD-I designer can use sound.

Chapter 7 The CD-I system
Looks at the impact of CD-I hardware and disc format on CD-I design.

Chapter 8 The design process
Takes the reader through the early stages of CD-I design, concept and treatment.

Chapter 9 Design analysis and prototyping
Moves the design process along to the stage where each creative and technical problem identified earlier in development must be analysed and a firm solution found.

Chapter 10 Detailed design
Continues the process outlined in the previous chapter to the stage where a title is ready to go into production.

Finally, the Appendix provides charts to help designers calculate the disc size of the material they plan to use, Further Reading provides a bibliography on all the subjects covered in this book, and the Glossary defines the technical terms used.

Preface

Contents

Contents

Contents

Contents

Contents

1

The potential of CD-I

Introduction

CD-I design is a new and expanding art. The titles that already exist only begin to reveal the possibilities of the medium. The advent of CD-I has added to and extended the possibilities of multimedia (explained in more detail below). Some early multimedia systems used amalgams of existing technologies, such as analogue video and sound – allowing the viewer to control a video cassette recorder or a slide projector by using a computer program, for example. Others used digital storage media such as laser discs or compact discs, but required separate computers and programs to control the player.

CD-I brings varied media together and stores them in the same way, as digital data on a compact disc, and the player and controller are the same device. It is not a technology added to a personal computer as an afterthought but is a fully fledged entertainment and information system in its own right. It conforms to a standard agreed by many electronics companies and has the full backing of its developers, Philips and Sony.

These revolutionary technical aspects of CD-I mean that it is very important to understand the medium before you can begin to design for it. You need to understand the potential of multimedia in general and of CD-I in particular. And you need to understand how CD-I works – how much data you can get on a disc, how it gets from the disc to the player, and so on.

With this in mind, this chapter discusses how CD-I relates to multimedia, explores some of its possibilities and looks at the opportunities for programme development. As with the whole of this book, you should always remember that nothing said here is exhaustive or definitive. CD-I is an open medium – there are few limits or boundaries, and there is always room for novelty and innovation.

The medium in context

Multimedia and hypermedia

CD-I is the most recent and technically sophisticated form of **multimedia**. It is also the first multimedia technology to aim at a mass audience.

'Multimedia' is a word devised to describe the blending of different media – visual, audio, graphic and computer – into a single experience. The opportunities this offers to designers and producers in areas from entertainment to education are only just beginning to be understood and tapped. Multimedia is turning into one of the fastest-growing areas of media production, and established producers of traditional information, such as encyclopaedias, films, television programmes and computer games, are exploring its possibilities. It has already generated a large literature and enormous interest.

Hypermedia developed alongside multimedia. It emphasizes the ability to access a wide range of information and data through a single interface, and to move from one point of information to another in a nonlinear fashion – you don't have to look through the preceding material to get to a point of interest. It exploits the ability of computers to access different types of information quickly, in a multimedia context. Information from separate video, audio and text databases can be integrated using a hypermedia approach. For example, a CD-I encyclopaedia would make it possible for you to look up an entry, read text about it, look at pictures, and hear music and spoken commentary. The choice of what you see or hear is yours.

Although hypermedia was first defined by the computer visionary Ted Nelson in the 1960s, it has become familiar to many personal computer users in the form of Apple's Hyper-Card program for Macintosh computers. This can be used to create and access graphics, sound and text held in files arranged as cross-referenced stacks of cards.

Compact discs and laser discs

Digital read-only media quickly became a focus for the development of new ways of storing and using information. Philips' LaserVision laser disc system formed the basis of several early multimedia systems, using computers to control the flow of images from the disc to provide an exciting interactive form of education or training.

In 1982 CD-DA (the standard audio compact disc technology) was launched. It quickly became clear that compact discs were a useful medium for distributing computer information. Such a 'CD-ROM' (CD read-only memory) is capable of storing up to 650 megabytes of data, many times the capacity of most personal computers. However, no single standard has emerged for storing or retrieving data from CD-ROMs. The CD-ROM XA (extended architecture) standard, defined by Philips, Sony and Microsoft, provides a firm standard and a link between the old CD-ROM and the much more useful CD-I.

Interactivity

This raises the topic of interactivity. More is said about this later on (especially in Chapters 3 and 4), but for the time being it is enough to note that interactivity involves a two-way communication between the users and the machine. The machine responds to their choices, and they in turn are stimulated by the machine to make more choices. Many devices, from banks' automated teller machines to teletext systems, are interactive to some degree.

Interactive video systems have been available before but most have been cumbersome and difficult to use, although some represent a triumph of design and programming ingenuity over the limitations of technology. CD-I represents interactivity in its neatest, most efficient and friendliest form. A huge range of devices – remote control, touchscreen, joystick, trackball – can be used to set up a dialogue with the CD-I player. Designers can choose the one that is appropriate for their audience and subject.

In short, CD-I takes the best of multimedia, hypermedia and interactivity and puts them all on a 12-cm compact disc that can be 'read' on a standard player anywhere in the world.

Interactive television

For this new technology, CD-I manufacturers have coined the phrase **interactive television**. By this they indicate the role they see for the new medium. It is not a computer, nor is it an intimidating technical device, but a development of an existing technology. Anyone who recognizes a television set, a TV/video remote control, a CD player or a video recorder will not feel uncomfortable with a CD-I player. There are about 1.2 billion television sets in existence in the world. This is the size of the market that is the potential audience for CD-I discs and players.

But this is not the end of the story. Although the idea of interactive television is central to the new product, there will be other types of interface. A variety of control devices allow the product to be adapted to more specific markets. Touchscreens, for example, are ideally suited to point-of-sale applications, while joysticks or trackball devices suit children's programmes. CD-I players too can vary, although there is a base specification to which all players must conform. But already , for example, there are players resembling personal stereos, for CD-I titles that require to be transportable (such as guides and maps).

Standards

Designers of CD-I titles can be sure that they are working with a standard medium and a mature technology. The CD-I standard specification ensures that throughout the world their product can be played on any CD-I player – or audio CD player, or computer CD-ROM drive fitted with an additional decoder. The full details of the standard are contained in the Green Book, which is the official version of the standard and was first published in 1987.

This is the field in which the designer of CD-I discs is working, an area that offers astonishing possibilities. The creation of titles and applications for consumer, education, training and commercial markets is wide open for the inventive and the innovative.

Multimedia and interactive design

What material can you use?

What kinds of information can a CD-I title use? The following are only a few ideas to illustrate the range of this medium:

Images

- Graphics: logos, credits, menu screens, titles, in-between screens . . .
- Text: explanations, documentation, help, instructions, directions, running commentary, dictionaries, encyclopaedias, glossaries, indexes, contents, footnotes, glosses, translations, novels, programme notes, biographical details, historical background, poetry . . .
- Moving natural pictures: documentary, drama, demonstration, entertainment, titillation, natural history, unnatural history, historical footage, home movies, Hollywood movies – in full screen or part-screens (windows), full motion or partial motion (page 77) . . .
- Still pictures: scenes, artworks, products, postcards, cutaway views, subtle montages, rapid alternations, stroboscopic fluctuations, family snaps, historical evidence . . .
- Animation: chases, captures, technical instruction, entertainment, humour, education, impossible actions and surreal situations, martial arts . . .

Sounds

- Music: background, foreground, attention-grabbing at a point of sale or information, musical analysis and instruction, forlorn or frantic, New World or old hat, hip-hop or hapless, jazz or Jesus, muzak or mazurkas . . .

- Speech: voice-overs, system instructions, programme instructions, help, sales pitches, dramatic recitations, poetry, incantations, whisperings, hypnotic suggestions, ruthless asides, snide rejoinders, pleas, prayers, Shakespeare and jokes . . .

- Sound: confirmations of actions, indications of change, signals of success or failure, disguises for seek times, prompts for action, humorous effects, reassuring backgrounds, worrying accompaniments, alarms, traffic, aeroplanes, warfare, animals, bodily functions, yawns, cries, guffaws, the wind, boots

in the snow, sawing logs, eating crisps, scratching your scalp, blowing your nose, doorbells, telephones, canned laughter, automated applause . . .

One of the beauties of CD-I compared with other media is that any information that can be captured or recorded in a digital format can be used in any title you produce. Because each piece of data – each snatch of sound or sequence of images – is stored separately, it can be combined and recombined according to the requirements of the programme. So a cheering clown in a children's quiz need be stored only once, to be called on each time the right answer is chosen, or at the end of each round, or when the programme is turned on, or even to attract parents' attention when the demonstration is playing in the toy shop.

How should you use this material?

Knowing what can be put on a CD-I disc is only half of the equation. To design a programme that uses the medium to its best advantage, you need to know:

Ways of using your source material

These are easily stated: you can use the interactivity of CD-I to combine all the materials mentioned above in any way you choose.

The fundamentals of interactive design

To make the best use of CD-I you need to have an idea of the issues involved in interactive multimedia design before you begin. Interactive television is a new experience not only for users but for producers as well, and careful thought is needed to ensure that your programme design works the way you want it to.

What you can get on a disc

Getting the right amount of material onto the disc is an important part of CD-I design. Is there room to scale up your plans? Or do you, on the contrary, need to rethink them to squeeze all your material onto one disc? This must be established as the design for the programme begins to take shape.

How best to use the disc space

You can juggle different qualities and types of material on the disc so that you use the space available to best advantage, and users aren't kept waiting while the player struggles to find the soundtrack for the images being played. Chapter 7 will give you some ideas. A summary table is included in the Appendix at the end of the book.

How to specify for others

Finally, you need to be able to communicate the fruits of your thinking to the people who matter – from producers of videotape segments, through graphic artists, to the programmers responsible for writing the control program.

What kind of CD-I programme could you produce?

Almost every category of entertainment and information devised can be produced for CD-I: television programmes, films, videos, books, audiotapes – the list is endless. CD-I is a suitable medium for bringing information into almost any environment – home and factory, classroom and shopping mall.

- Education and training
  Do-it-yourself – from woodwork to car maintenance, via plumbing.
  Home learning: history, languages, accountancy, interactively enriched.
- Interactive training
  Learning to drive, operate a jib crane, arrange things in warehouses, take photographs that come out well.

- Reference books

  Illustrated talking encyclopaedias, dictionaries, multilingual dictionaries, phrase-books, learned journals, classical texts (taken apart and explored).

- Educational forums

  Seminars, lectures, discussions and debates.

- Interpretation

  Music, paintings, sculptures – examined and analysed, compared and criticized.

- Physical fitness

  Bespoke workouts, physiotherapy, relaxation exercises.

• Entertainment

Drama: stop the action and find out more about a character or situation.

Music: pop 'videos', interactive music (try different arrangements and instruments, and write it yourself), soothing pictorial accompaniments to airport lounge music.

Children's programmes: join-in nursery rhymes, cartoons for colouring-in, mazes, quizzes – television as surrogate parent.

Action games: convincing simulations of anything from a bombing run to the Le Mans 24-hour Grand Prix.

Adventure games: see the world, destroy it and start all over again.

Activity simulation: play golf without leaving your seat, follow a team through the football championships.

- Commercial

   Point of sale: the most articulate sales pitches (which tell you what you actually need to buy, and why).

   Information point: everything you need to know about trains and planes, shops and hotels.

   Mail-order catalogues: demonstrate products and make it easier for viewers to find what they want.

   Parts catalogues: an end to microfiche – see what the part looks like before you go and search for it on the shelf.

   Procedures: introduce new staff to the way things are done.

   Security/identity checks: compare faces and signatures with accurate pictures of their originals.

Manuals and guides: with animated examples, quick cross-referencing, optional in-depth explanations.

Travel guides: see your destination, have a look around the sights, even peek into the hotel bedroom before you buy your ticket.

Museum and gallery guides: brush up on what you want to see and where you can find it.

Conclusion

CD-I is an exciting new medium for entertainment and information programmes. It differs from earlier multimedia technologies by being available in a fully integrated and standard form. Discs are playable on any dedicated CD-I player, or converted CD audio player, around the world, and are physically the same as CD audio discs.

The versatility of this all-digital medium is such that any kind of information that can be captured in digital format, whether sound, still or moving image, cartoon or photograph, can be included on the disc. The range of possibilities means that CD-I can be used as the basis for almost any programme.

Designing CD-I programmes, however, needs three things: a clear and original programme idea, an understanding of how to make it work as an interactive programme, and an understanding of the technical features of the CD-I system to make sure that the proposed title makes the best use of them.

2

CD-I as a medium

Introduction

This chapter looks at the first part of CD-I design: identifying a market opportunity and developing an initial concept that will satisfy it. It considers the market for CD-I discs – who is going to buy or commission CD-I discs, what they will want to see, where and how they will make the buying decision. The chapter concludes with some tips for effective CD-I development and design during the initial stages.

Markets and opportunities for CD-I

The versatility of CD-I means that it will be used in a great many settings, all of which offer different opportunities for the producer of programmes. It is applicable to both the consumer and the business markets, among others.

Custom CD-I for business

There are many ways in which business will want to use CD-I, as suggested in the previous chapter. Developing a programme concept for in-house use by a client's own staff could be quite straightforward: a chain of department stores could commission a programme to train new employees in company procedures and customer service, and the material to be included would flow from that, incorporating everything from processing credit-card transactions to returning faulty stock.

Interactive training is a well-established medium and translates well to the CD-I environment, which simply makes it easier to use and hence more effective. But this is not the only way in which businesses will want to use CD-I: promotional programmes, catalogues, guides and so on may benefit from a more consumer-oriented design approach. After all, with CD-I business programmes need not be staid or boring, and a low budget doesn't preclude the use of many CD-I effects.

Mass-market and consumer CD-I

Consumer interactive television is a less well-established medium. It's likely that many early CD-I programmes will translate an existing television programme into the new medium – for example, turning an established quiz show into a family game – and so the challenge for the designer will be to reinterpret existing linear audio-visual material for the interactive CD-I environment.

The audience

To begin with, the audience will be unfamiliar with CD-I. So try to use the medium efficiently and consistently in order to give them as much help as possible.

One of the first tasks in designing a CD-I is to profile the audience. Will they be professionals, using a training CD-I at their desk? Or a family, watching as a group in their living-room? The answers to these questions affect more than the content of the disc. They also have an important impact on its look and feel, the screen design, the production values, size of text and so on. Try to analyse what the audience's expectations will be, in terms of production values and style of presentation. How much help and reassurance will they need when using the CD-I and what will their attitude to it be?

You should aim to be specific about your audience. Taking a scattergun approach and aiming at all audiences will please none. And while two different audiences may share an interest in a subject, they may need to have the subject treated in completely different ways if they are to be engaged by it. One group may expect a chatty, informal approach and be put off by a 'neutral' tone, and vice versa.

Remember that accurate targeting of the audience applies to multilingual presentations as well. The CD-I audio track may be used to present several different language tracks. Check carefully that the pace and style of accompanying visual material are equally appropriate for each of the national audiences.

In the early years, it is likely that many CD-I products will be adaptations. In audience terms, this can be helpful. The audience will already be familiar with the underlying concept of the programme, at least. Your job will be to give them a reason to buy the product in CD-I form. But even if the material is familiar, interpreting it for CD-I needs considerable thought. A primer for schoolchildren will need to be easier to use than a reference work for adults – or are you going to assume the presence of an adult operating the system?

CD-I on location

Whatever the programme under development, where and how it will be used are two of the most important factors to consider when designing it, and they provide clues to the needs of its audience. This information tells you who the users are likely to be, how familiar you can expect them to be with the very idea of CD-I and other interactive media, how long they are likely to spend using the disc in one session, and whether they are likely to use the disc just once or more frequently.

This in turn gives clues about how the content of the disc could be structured, and the approach the programme could take to the information it conveys. In a town guide for tourists it might be appropriate to wish users a pleasant visit at the beginning of every session, because each user is likely to encounter the programme only a very few times. But a cheery greeting sequence would be inappropriate on a disc that one person will constantly refer to.

Conversely, the staff training disc doesn't really need a rolling demonstration that shows the disc off when it's not being used – but the television-derived quiz game would benefit from such a demonstration to attract attention in a store or rental outlet.

Estimating production costs

Both client and producer are going to be concerned about the cost of CD-I production. You may need to show, for example, that producing a CD-I programme will be as cost-effective as a traditional video, catalogue, or book.

Production costs can vary as widely as the programmes themselves. A programme that uses pre-existing sound and images belonging to the client will cost a lot less than one that requires full motion video to be shot in an expensive location with expensive actors. But the costs of source material are also a major consideration; and the post-production cost of turning material into a CD-I programme will also be a variable factor.

The budget for the programme will obviously influence its design. CD-I has an advantage in that multimedia trickery can make very effective use of the material but, just as with traditional television production, it is probably better to scale the production to the budget than to attempt more than the money allows and achieve an unconvincing result.

Developing concepts and material

Programmes originate in lots of different ways but successful ones have in common an imaginative concept, a firm appeal to the market and strong source material.

The concept usually comes first. This could be a brilliant idea for a game or simulation. It could be a training programme, the objectives of which might have been stated first. Sometimes the concept comes from a firm grasp of a target audience and its needs: Do-It-Yourself enthusiasts, company directors, children from five to eight, golfers . . . . Or it may be that pre-existing collections of film or audio materials spark off ideas for CD-I titles.

Initial concept

There are well-established ways of developing an initial concept, in broadcasting, film-making and the audio and software industries. Your brief may be quite clear from the start, and suggest the entire structure for the programme by listing everything that must be included. You may be asked to turn an existing product into interactive format; this is a conceptually simple process for reference works from encyclopaedias to mail-order catalogues, and for media forms that are already interactive, such as quizzes and games. Their structure transfers neatly to CD-I and provides a simple and coherent framework for the programme, although that still leaves detailed aspects of design and interaction to be created.

With other programmes this process may require a lot more thought, if the CD-I is to be different from any other medium that could use the same material, and is to be more exciting. A CD-I producer might simply be asked to make a CD-I 'about', for example, a rock group. There are a huge number of items that could be included in such a programme, and many different ways to organize them:

- The *Greatest Hits* video collection could be played straight through, just like a conventional video.
- The user could select songs individually.
- He or she could choose to read a lyric sheet or the music rather than watch the promotional film featuring the band.
- Songs could be selected according to speed, volume, theme or any other feature that could be indexed.
- Interviews with the band and reviews of their concerts and albums could also be featured.

In fact the first challenge to the CD-I programme designer is to build an interactive framework for the material. Try to make it integral to the whole design, so that it is impossible to imagine the finished CD-I without it. Interactivity should not just be tacked on later as part of the detailed design. If the interactive element is identified as a key component of the basic concept, it will be far stronger and far more involving for the audience.

The type of interaction you intend to build in is important to the programme structure and its style. If you intend viewers to progress through the disc at their own pace, the structure will be more like that of a traditional television programme, computer-based training game or leisure computer game. Different sections could be indexed by keywords so that they can be accessed individually, hypermedia-style.

Because CD-I design involves new processes and technology, it may be worth checking that your existing procedures for developing concepts into programme outlines get input from everyone who should be involved. Bringing in the whole team, from producer to programmer, is a good idea, and a brainstorming session may result in a host of ideas, which can then be sifted. Although many ideas will be rejected, you can be confident that you have involved someone with a knowledge of each aspect of CD-I design and made it less likely that you will have missed any good ideas that make the best of CD-I and your material and facilities.

Will your programme idea work as interactive television?

Is there a need in the market for this title?

Could an interactive programme add value to the versions of this material that exist in traditional media?

Do you have the rights to the material, characters or performers you want to feature?

Does the programme involve the viewers?

Is the programme concept technically feasible?

Can the programme be made within the budget available?

Is the programme easy for the intended audience to use?

Reviewing existing material

A thorough knowledge of existing material is essential. Is there enough suitable material to avoid the need to record any extra live action or sound? If more material needs to be recorded, are all the participants and the set available? Does the material all work well together, and are there any obvious continuity problems? This is a major pitfall of working with existing material. For example, a CD-I of conjurer's tricks could cause more amazement than intended if the assistant changes shape, size or sex halfway through.

Furthermore, looking at the material you already have – the television programme or film on which a CD-I is to be based, for example – could give you a strong idea of ways to structure the CD-I version.

Other resources

Designing a CD-I raises all kinds of creative, financial and technical questions. If you're entirely new to interactive tele vision design it may well save money in the long term to budget initially for some consultancy. A CD-I expert is aware of both the potential and the constraints of the medium. He or she can assess your design and spot potential problems. Such a consultant may be able to save you significant sums of money by giving expert advice at an early stage.

If you're planning to use an established CD-I production facility there should be someone on hand to provide this type of advice and ensure that your efforts aren't wasted because you're attempting the impossible, or not getting the best result with your material through a simple oversight. Philips has a CD-I facility available for outside use and it's likely that many existing video facilities houses will develop CD-I expertise.

Usability checklist
Have you got ideas for this programme already?
Will they work with the material?
How interactive will the programme be?
How many times will the viewer use it?
Does this make a difference?
What will happen if something goes wrong?
Can the user get help?
Are there shortcuts for the user?
If the user switches off in the middle, what happens?

Developing the programme idea

By this stage you may have a clear idea of what you intend to do, how your programme will be structured and what material it will feature. Or you may still be developing it from an initial concept, drawing together several threads that will eventually become the programme. Look at the ideas from various angles in turn. Which of them will appeal most, be most instructive, be hardest or easiest to follow? Is there enough here to sustain a disc? Or too much?

Then look at the suggestions from a broad design point of view – not in detail at this stage. It's a good idea to build in a review around this point and to summarize the concepts that you have developed so far.

Because all interactive techniques are relatively new, there are no real scripting standards as there are in film- and video-making or in programming. The degree to which you need to be explicit will vary, depending on who your script is aimed at. If you are working closely with a creative team, it may be quite loosely defined. If it is going to an outside production facility it will have to be far more detailed to ensure that your intentions and ideas are clearly communicated.

'But I know what I don't like . . . '

People sometimes find it hard to convey the 'look and feel' they are aiming for. One useful technique to get over this is to hold a viewing session with your coworkers. Look at other CD-Is or related multimedia presentations. Find out which ones people like or dislike and encourage a group discussion on their reasons for their reactions. It is far easier to establish a common vocabulary when you are both referring to the same examples. In this way, the designers can get a better feel for what the client wants.

Structure

Designers need to think hard about how events will be sequenced and material structured on the disc. A dull or muddled disc structure can spoil a programme, even if the sounds and images are excellent.

Once the structure as the user will see it has been established, the underlying structure of the disc itself will need to be developed: the programme material needs to be mapped into the disc. This technical process is discussed in greater detail in Chapter 7, but it has implications for programme design that need to be considered at the start of the design process.

If your design allows the viewer to jump to any point in the programme at random, it may sometimes take a short but perceptible time (around a second) for the disc to retrieve the requested programme segment. You will need to devise ways to disguise this delay or minimize it. Try to group associated items so that the time taken to search for and display an item is minimized. If such delays are going to occur, think about incorporating lively, consistent screens that cover these seams, so that to the user they look like the introduction to the next section.

Questions of structure are closely related to interactivity and the way your CD-I will be used by its audience. If, for instance, you have a menu that is accessible from every part of your programme, think about positioning it where it is most quickly found rather than putting it at the beginning of the disc. Or consider having two or three identical menu screens placed across the disc, to cut down access time. Once the structure has been decided on, programmers, authors or designers can begin to implement it.

Modules

Unless you have an unlimited budget you will need to maximize the resources available to you. Splitting the design into modules that can share some of the same resources is one way to reduce the burden; it enables you to recycle as much of your work as possible.

Your programmers will be able to use some resources that have already been developed. For instance, Philips has a number of off-the-shelf 'engines' that can be used to help with tasks such as animation. Chapter 9 discusses modules and software engines in more detail.

At this stage you might want to develop a sample module or two, as part of an iterative development process, to see what they're going to look like.

Branding

If a CD-I is to get attention, it is important that the overall impression be attractive, with carefully chosen colours and graphics. If you are planning a series, you will want consumers to think of your CD-I series as a brand that they will buy, enjoy and look for again. The design concepts must therefore be consistent. The packaging, any accompanying booklet, the screens and the general visual style must all reinforce the message about what to expect from the series and each CD-I that is part of it. This applies equally to training and corporate CD-I packages.

The interface

Design of the user interface is covered in detail in the next two chapters. You must take into account how near the set people are likely to sit when using your particular programme and how many people at a time are likely to use it. These factors will vary greatly and will have a major effect on the interface. Also, match the interface to the user and don't forget that an imaginative interface may help to sell a consumer product.

Scheduling

It's clear that, for good time management, several processes must run simultaneously: graphic design, content design, script-writing, filming, recording, research and programming will probably all need to be scheduled to take place simultaneously. The final chapters of this book look in more detail at this multi-disciplinary design process and the different ways in which it can be managed successfully.

Design tips

Be interactive, not hyperactive

Do not be tempted to put in lots of interactivity just for the sake of it. Users may find it frustrating, irritating and eventually boring, if they feel they cannot get anywhere or do anything without jumping through a series of hoops. This is especially the case in home entertainment CD-Is, which may be played many times over.

Don't over-design

CD-I is an extremely versatile medium: all kinds of things are now possible that once were not. However, that does not mean that they all have to appear in your CD-I. Remember, you are designing to inform or entertain your audience, not to impress your colleagues. You will be better off spending time getting a reasonably simple design absolutely right than a complex one half right. 'Remember to put your audience's needs first. It's too easy to design to impress other designers, to go for awards for technical brilliance, and produce something that is too complicated,' commented one designer.

Design for sequels

What happens if your CD-I is a huge success? If you are likely to embark on a follow-up, or another title in the sequence, bear this in mind when you start the initial design. Try to design as much as possible – scripts, interactivity, screen designs, material, programming code – so that it can be reused with fresh material. Using modular designs (Chapter 9) may help here. This will greatly reduce development costs on follow-up discs, and may mean greater profits or extra resources to spend on originating new material.

Test all the time

To begin with, develop sample modules on paper. If possible, cut test discs and get novice users to try them out. Otherwise, use development software (called prototyping tools – Chapter 9) to try out your title. Check discs that will need to be converted from other television standards as early as possible and test the production routes if you can.

Teamwork is vital

Every element of CD-I is interdependent. Everyone in the design team must not only work together but communicate regularly as well.

Make tradeoffs

The CD-I disc capacity is large but not infinite. Work out early on how much video, audio, graphics and other material you would like, and how much space each is going to take up. Then if you find that you've got too much, you can decide how to compromise, trading off technical and design decisions against each other.

Hook the audience

Hook the audience's attention from the word 'go'. Put your best work up front and design a rolling demo or a demo section.

Design it to last

Remember, the audience may replay the CD-I dozens or even hundreds of times. Sustain their interest by means of challenges, rewards and humour, and by offering them different ways to use the disc, such as short cuts.

Create a believable world

Create a believable world by extending the design to the packing, cover and any text material. Make sure that icons and interfaces don't create distance between the programme and the audience.

Link presentation to content

Although this is a tenet of many design disciplines, it's worth restating for CD-I, where the nature of the medium may make it difficult to see what is the appropriate presentation for a particular subject. One designer says that you know that the design is right when viewers find the programme transparent and the medium doesn't get in the way of their enjoyment.

Make the meanings of icons obvious

Use icons only if they are appropriate for the programme environment. Make their meaning clear and obvious: otherwise give clear directions and do not burden the user with too many instructions to be remembered.

Conclusion

Effective design for CD-I requires one more element than design for conventional linear media: a structure for interaction. It's important to have an exciting and well-developed pro-gramme concept, whether it's entirely new or transferred from another medium.

The special features of CD-I need to be considered from the start of the design process. It's not enough to say 'Let's produce an interactive X'; using CD-I must add real value to the pro-gramme, providing something that no other medium can, whether it's ease of use and access to information, a more exciting audio-visual experience, or the potential for the viewer to order the material to his or her own liking. Thinking about who the audience for a disc will be should play an important part in the design process.

Once the theme and purpose of the disc are established you need to consider three things in more detail. The programme content of the disc must be fully developed, the way the user interacts with it must be finalized, and the way the programme will be stored on the disc must be considered in case poor disc layout affects the viewing experience.

3

Designing for interaction

Introduction

This chapter outlines some of the design considerations that apply particularly to CD-I titles, and explains how to develop your concept to ensure that they are factored in from the start of the design process. Its emphasis is on including interactivity from the start, and on working carefully from the initial concept to a more detailed design. Hints on developing the programme concept through the use of flowcharts and diagrams, storyboards and scripts are given, along with examples.

Interactivity is not the only means to interest viewers in CD-I programmes. Traditional film, television and drama rely on excitement and involvement to hold the audience and this should not be lost sight of in the pursuit of interactivity.

Flowcharts

Interactivity is multidimensional, paper is not. For this reason, trying to describe interactivity on paper always causes problems. In particular, flowcharts of interactive sequences can become complicated and difficult to use. Yet documentation of the programme's structure is necessary. Some kind of diagram is essen-

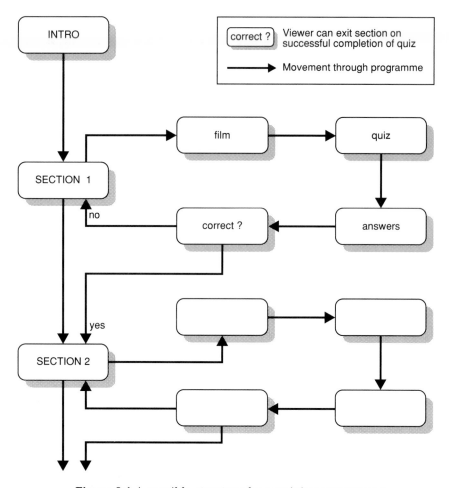

Figure 3.1 A possible structure for a training programme.

tial to enable you to develop the structure of your programme. Unlike a linear television programme, the structure of interactive programmes is explicit and visible to the viewer, who has to make choices in order to use it. If the viewer doesn't understand where to go next, how to start watching the programme, how to finish or how to get to the particular piece of information he or she wants to see, your programme will not be a success.

The **flowchart** is a good way to get ideas for the backbone structure of the programme onto paper, and to consider different ways of organizing the structure of the programme. Flowcharts are also a traditional way of designing the structure of

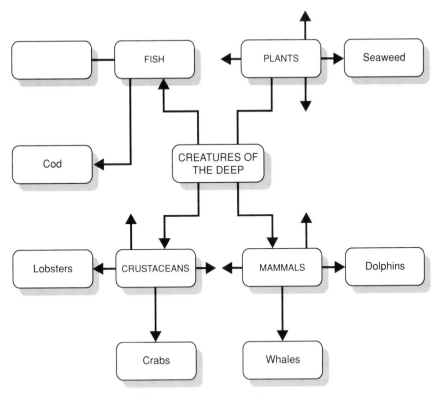

Figure 3.2 A possible structure for an informational programme.

computer programs, so for the benefit of your programmers you will probably need to describe your programme idea in this format at some point.

Figures 3.1 and 3.2 show how a designer can use a flowchart to work out the structure of two different kinds of programme. In the training disc (Figure 3.1), it's important that viewers go through every section, and can leave the disc only when they've completed a section by passing a test on the information it contains.

The *Creatures of the Deep* wildlife programme (Figure 3.2) takes a hypermedia format, and the information is stored in several layers, like a family tree. Viewers can reach the creature they want to see by requesting more and more specific information; but you could add other paths through the data to specify creatures by what they ate or where they could be found and so on, thus taking the viewer to the same information about each creature. Figure 3. 3 shows how this could work.

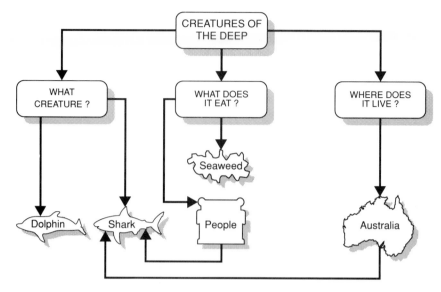

Figure 3.3 Alternative paths through the programme of Figure 3.2.

But for a medium as potentially dense as CD-I, a flowchart that does more than name each module or section of the programme is likely to become unwieldy and will not achieve your goal of creating a simple diagrammatic representation of your programme idea. Some designers solve this problem by making a flowchart of the main sections of the disc, to show the main moves the user can make around the programme, and using other techniques to describe the details of each section. These can include:

Numbering

Give each 'level' a number, where the top, or menu, level of one particular option is 1.0, the next one down is 1.1, the next 1.2 and so on. This works well for hypermedia formats and reference works, where the user is likely to be looking for a small piece of information rather than wanting to watch an entire programme.

Narrative

Narrative format, in which the main interaction is described, followed by lists of the questions or options. The interactive logic is described rather than fully illustrated. This might work well for the training disc example above, where the user's progress through the disc is highly managed.

Codes

Coding systems, in which differently shaped boxes or screens represent different interactions or types of data, could be used. You might use different boxes to represent audio-visual sequences with no interaction, places where the viewer has to make a choice, places where the viewer can leave the programme, and so on. A good system would make a flowchart easier to read, but take care to be consistent and make sure that everyone who needs to read the chart can understand the codes. Figure 3.4 shows the training disc flowchart of Figure 3.1, modified using a set of codes to make it simpler to use.

Mixed flowcharts

A flowchart can be limited to showing the main interactive options, and using some narrative, to prevent it from becoming too complex.

Computer-based modelling

One option is to abandon paper altogether and use software of the HyperCard type (page 3) to model the interactivity. For experienced CD-I designers, this may be just as easy as using pencil and paper, and it might replace even the most tentative initial doodles. This approach allows the interactive structure and the users' options to be modelled realistically, and means that complex interactions and paths through the programme aren't a confusing jumble of arrows and notes on your piece of paper.

Scripts

A programme can be represented in outline by means of a traditional script, using a narrative format, at the start of each section, followed by a modified script, with three columns

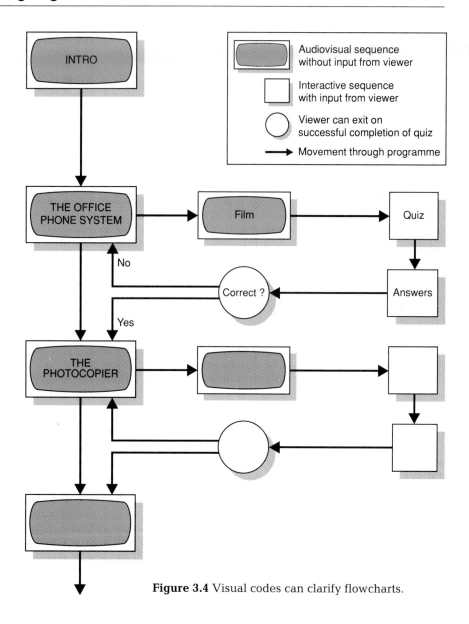

Figure 3.4 Visual codes can clarify flowcharts.

giving details of sound, visuals and interactivity. This, however, is a much less useful way to proceed at this stage, even for those programmes that offer the viewer a highly managed path. A visual representation of the programme's structure is likely to be much more adaptable and easy to use.

Scripting and storyboarding

Scripts and storyboards are two ways of further developing your concept. In the early stages of the design it can be useful to emphasize storyboarding rather than script-writing. This is because CD-I is a visual medium and using a storyboard to explain your ideas can encourage you to think visually from the beginning. A good approach is to attach storyboards for each section to the relevant place on the flowchart you have already made.

The storyboard is also important when the CD-I involves concepts that are hard to explain in words. Like the script, it helps to make the design concept explicit. It is part of the process of communication and of creating a common vision among the design team.

If your CD-I involves the production of new audio-visual material, you will of course want to develop and express the ideas for this by using a storyboard, just as you would for any other type of film or video. It's an ideal way to communicate your visual requirements to the director and studio staff. Other uses for storyboards in CD-I include screen design and interaction design, to show how you want your programme to appear. You may, for example, want to show how a user will be presented with choices, the visual effect of making that choice, and the way the programme moves on to displaying the requested information. The short storyboard in Figure 3.5 shows how a training disc moves from one module to the next.

Programmes that involve new material will obviously also require a script for any film or animation sequences. If you're reusing film or computer-generated animation that has already been made, you might wish to record a new voice-over for the CD-I version. That too will require a script to be written for the use of the performer, director, producer and studio staff.

Although scripts are vital for sections of CD-I programmes, they are not the best way of communicating everything that will be part of the programme, and should be used in conjunction with flowcharts and storyboards as just one of several production tools. When you write the dialogue or voice-over, you

Figure 3.5 Storyboards are a good way to demonstrate interactive sequences.

shouldn't rely on the viewer having seen any other sections of the programme, unless it has been structured so that they are bound to have done so. Continuity is a problematic area in interactive media.

Do you need a scriptwriter?

Dialogue for audio-visual sections of a programme is only one part of a potential CD-I script. Any on-screen text also needs to be carefully written. If there isn't a full-time writer on your team, someone needs to take responsibility for the text. Text on screen has to work hard – it must be concise and informative but must not patronize the viewer. It should also be factually accurate, correctly spelt and easy to understand.

So you may wish to engage a professional scriptwriter. It is very much up to you to decide at which point to bring them in. You may decide against using someone for the whole script, and instead employ them for a day or two to comment on the words that have been written. Writing will need to take place alongside screen design, so that the words fit into the space allocated.

Developing the script and storyboard

Scripts and storyboards will go through several versions during programme development. In the beginning they are very general documents that give people a broad idea of the shape and style of the programme. Through review, feedback and redrafting, they become successively more detailed. However, at the start it is usual to give a thorough treatment of a short module so that the design team can use it to judge whether the ideas are likely to work in practice. The final chapters of this book look at this process in more detail.

As the script/storyboard becomes more detailed, it will include key images and sequences, and information on technical matters such as the use of planes for different layers of video information. It may also have a running comment on the CD-I's **context** at any given point. The context is the state of the CD-I player and the information it is displaying. It is likely to include: how many planes are in use; what kind of media are loaded; whether the cursor is being displayed; whether sound and video memory maps are loaded; and what runtime files are loaded.

Developing the design

Demonstrations

While designing the programme itself you also need to consider the demonstration. The precise form this will take depends on the type of programme being made.

If the CD-I is to be demonstrated in a shop, then the demonstration must show the most exciting and attractive aspects of the CD-I. So it is worth including a rolling demo on the disc that will show the CD-I programme at its best. Take a look at film and television programme trailers and the way they show the most dramatic, scariest or funniest moments in the film.

Point-of-sale and point-of-information CD-I discs will need to include a demonstration that can be played whenever the system is not being used. It will need to be carefully designed so that it shows what the disc can do but is easy to interrupt and doesn't deter anyone from approaching the system. You will have to choose between a short, instructive demo to encourage people to interact with the system, and a longer demo that shows off more of the disc. See the section below on automated presentations (page 41).

If it is not possible to include a rolling demo, you may want to make sure that one section of the disc is specifically designed to be a showpiece for the disc as a whole.

Involvement

When you have hooked the audience, you need to involve them. They must want to buy the CD-I and take it home in order to see more.

To do this, you need to create a believable world that draws people in. This world can be totally fantastic but it must maintain its own credibility. It does this by creating an illusion and then consistently reinforcing it. Sudden jumps to graphics screens in a different style or the intrusion of inappropriate audio effects can destroy the illusion and break audience involvement. They should not be able to spot the seams where you have stitched together audio, video and graphics.

Responsiveness

The response to the audience must be positive and encouraging – which does not necessarily mean patting them on the head. The programme could tease them for failing to spot a clue. But the response should be one that maintains their involvement, by encouraging, amusing or challenging them. A 'positive' response could just mean giving the user simple reassurance.

Remember too, to make your design as flexible as possible. If the CD-I has been designed for use in a family setting with viewers of both sexes and different ages, it must offer something to all of them.

The pages that follow discuss some of the methods that will help CD-I designers to fulfil these design goals and successfully launch CD-I titles in the marketplace.

Automated presentations

Automated presentations, such as point-of-sale information systems, have their own considerations:

- Will the programme run continuously?
- Who will use it?
- What will the lighting be?
- What about traffic flows round the player?
- Should it be interactive and how?
- How long should any information be displayed?
- Is the system likely to irritate anyone working near it, such as store staff?

The effect of an automatic multimedia information system on people who have to work with it may sound like an amusing issue, but it is worth thinking about. Your system will have an impact on someone's work environment and you should consider this. Workers have been known to vandalize irritating machines. The use of sound in point-of-sale and information systems can have a negative as well as a positive effect.

It is important to have an awareness of the current conventions used in broadcast television. These represent a kind of visual and aural 'language' to which people respond, even though they may not be aware of it. Look at current television programmes aimed at audiences similar to those that your CD-I is intended for. Take note of the way they present text, music, programme breaks, linking screens and so on. It is very likely that the expectations of your audience are moulded by these programmes.

Think carefully about what your audience knows already. If they are corporate users they may already have used interactive video. If they're children, they may be unfamiliar with CD-I but eager to explore and try out something new. Alternatively, they may be older, more conservative and a little more uncertain. If this is the case, it is even more important to be consistent and clear in the CD-I design.

Test pictures are very useful at the start of a CD-I programme. They allow the user to check that the colours are correctly displayed on the screen. A 'clean' finish is important too. Ensure that the CD- I player is left in a 'ready' state by your programme when it has finished.

Audience profile checklist

Who are they?
Where will they see your programme?
What do they expect to get from it?
What are you offering them?
What do they know already?
What is their attitude to CD-I likely to be?
What kind of 'look and feel' do they expect?
What kind of presentation will engage them?
What kind of presentation would they find off-putting?
How many times will they watch your programme?

Managing user interaction

Interactive media should not be anarchic media. While you should encourage the viewer or user to interact with the programme, you should not abdicate responsibility for it. The care you take in designing your programme will determine their enjoyment of it. The information still needs a structure, which it is your job to design, and the user needs help to navigate around the programme safely.

Consistency

It's vital to be consistent throughout the disc. If you're making a series of discs, they should be as consistent as their subject matter allows. You should aim for consistency of presentation, so that screens look reasonably similar from section to section. This is especially important when the user is likely to jump around the programme at will, as in an encyclopaedia. Creating a good interface and sticking with it is the best way to ensure programme consistency, which in turn is vital for ease of use; the next chapter discusses interface matters in more detail.

It's also important to make sure that there aren't any loose ends or difficult corners in the structure of the programme, such as a section that the viewer can't return to the main menu from.

Continuity

Continuity is a difficult but important area for interactive media. If you're working with old material, you must make sure that there aren't any gaping continuity problems that might disorient the viewer. Don't lose any characters, or suddenly change the voice giving the voice-over, without explanations.

For example, a children's CD-I might include several episodes in the life of a character. If she has a pet cat in some episodes and not in others, do you feature the episode where she acquires the cat? And if the sudden appearance of the cat is puzzling, can the viewer use the interactive facilities of CD-I to find out more about the cat and where it came from?

Pace

There is also a need to keep a steady pace in your demands on the viewer. Don't rush viewers into making choices, or blitz them with too much information in a short period. Try to balance the amount of information that is on each screen, so that the viewer can make a steady progress through the programme. Of course, with some types of CD-I, such as games, your demands on the viewer will be neither slow nor steady.

Maintaining an editorial perspective

When viewers are able to pick their own way through a programme, there is a danger that they may never see the programme you intended to make. For some CD-Is this won't be a problem, because the viewer's path has been carefully laid out. But some programme types need special attention to make sure that your editorial intentions are clear whatever way the viewer uses the programme.

Take the example of a CD-I documentary about the US Presidential elections. You might wish to feature voters talking about their opinions. In a linear documentary you could make sure that all viewers saw the balanced selection of opinions that you had chosen to highlight – unless, of course, time-shifting viewers fast-forwarded through that section on their VCR. But how can you arrange the speakers on your CD-I so that the viewer is guaranteed a full range of opinions? And should you bother to do so?

The most obvious arrangement would be to make each speaker a separate section, individually selected from a menu screen by the viewer. You could arrange the screen so that the speakers were explicitly divided on-screen into those in favour of a policy and those against it. You could arrange the programme so that opinions could only be heard as part of a matching pair of pro and con statements.

Political documentaries are only one example of a CD-I programme in which you might want to make sure that the viewer sees certain sections. In a Do-It-Yourself manual you might want to precede the electrical repairs section with a safety lesson on the correct way to wire a plug, and only allow the viewer to go any farther once he or she had shown that they knew how to do this correctly.

Conclusion

Once you have established the central concept for your CD-I programme, you should move on to designing its overall structure. Still considering important facts such as where the disc will be used and by whom, develop an overall structure appropriate to the information your programme will contain and the way you want to organize it.

Flowcharts, storyboards and scripts are useful tools for developing this stage of your design, and can be expanded and refined as you finalize your plans. Don't forget to include a demonstration section to promote the disc, where appropriate.

There are two main pitfalls to watch out for at this stage of the design. Don't make the mistake of thinking that the extra control that viewers can exercise over interactive media means that you don't have to design the structure of the programme carefully. If anything you have to take more care to ensure that the programme is easy for viewers to use while still carrying the ideas that you wished to communicate. The second mistake is to fail to involve viewers in the programme in the way that a traditional one would. Being able to manipulate a mouse and click a few buttons is no substitute for a good story, interesting and well-developed characters and well-presented information.

4

The interactive interface

Introduction

This chapter looks at the question of interface design in detail, considering the impact of screen design, sound output and user input on the usability of the programme. It is intended to provide ideas and suggestions for the concrete design of a programme interface, down to the design of each interaction. It builds on the last chapter's discussion of the development of an outline structure for an entire interactive programme.

What is the interface?

Is the interface:

- the screen?
- the text on the screen?
- a hotspot (an interactive area) on the screen?
- the cursor?
- the remote control?
- the concepts on the disc?
- the on/off switch?

The answer is that it can be any or all of these things, and indeed many others.

The interface is the point at which the system – screen, disc, remote control – meets the person using the system. It is the means by which that person gets information and if necessary sends a response back.

The computer world has devoted a great deal of effort to researching the way people interact with machinery, in order to develop computers that are easier to use. CD-I designers can learn from this research, which covers an incredibly wide area, from on/off switches to the colours of document icons. Although the demands of CD-I design and interactive television are somewhat different from those of the ordinary business computer application, the research is relevant in many ways and will help you to make some important design decisions, such as the choice of input device or how to manage the cursor.

With CD-I programmes it is important to remember that the interface can be fun. An exciting new input device, such as a glove or a gun, could even help to sell a consumer entertainment disc.

Input devices

CD-I supports a range of input devices and you will need to choose the best one for your programme, or to ensure that it works with a range of different devices. In some environments you may even be able to create a custom device to encourage users to interact with your programme. For instance, an interface for children might consist of a large pink trackball, with big yellow ears and whiskers. Its key qualities might be that it is cheerful, wipes clean and is unbreakable.

As well as considering the needs of your programme, you'll have to bear in mind the type of machine on which it is likely to be played. For example, you should not request keyboard input in a programme that is aimed at the domestic market, where many players will be 'base case' models without keyboards.

This might have an impact on the entire design of your programme; for example, an encyclopaedia for home use will need
a clever structure to enable users to get to individual definitions
quickly without too much typing. But a business user is more
likely to have a keyboard, or to be buying a CD-I system
specially for your custom programme, so it would then not be
unreasonable to demand that a keyboard be available for input.

However, all CD-I players are fitted with some form of X-Y
pointing device – that is, one that moves the cursor left–right
and up–down, in order to point to and select items on screen.

The remote control

This is one device that is currently unique to multimedia
formats. It allows the user to control both the player and the
programme. When it controls the player it works like the kind
of remote-control device familiar to everyone with a modern
television set or VCR. When controlling the programme it is
much more sophisticated, allowing the viewer to point at and

select on-screen items, which will usually be hotspots leading to another part of the programme. This combination unit is known in the CD-I world as the thumbstick. All consumer CD-I programmes should be made with this device in mind. It's easy to use and allows quite complex interactions. The user can control sliding-bar images on screen, such as volume controls, by repeatedly clicking a button or holding it down continuously.

The mouse

This pointing device is much more familiar to computer users than to the ordinary CD-I viewer. The physical movement made by the mouse clearly relates to the movement of the cursor on screen, and the user can move the cursor without needing to look at the mouse.

A mouse might be appropriate for business discs. However, it needs a flat surface to enable the ball at the bottom of the mouse to move freely. This limits its usefulness in situations where users are not likely to have a suitable flat surface, such as in the home, where a viewer might be watching a CD-I while sitting in an armchair. Most mice need to be connected to the system by a cable, and this too limits the situations in which they are likely to be useful. However, mice linked to the player by infra-red beam are available.

The CD-I mouse has two buttons, like the mouse used with most IBM-compatible computers, but unlike the single-buttoned Macintosh mouse. Understanding the different effects of the two buttons is often the most difficult part of using a mouse.

The joystick

The joystick is familiar to players of computer games and is possibly the most suitable device for game programmes, where constant, frequent and reasonably accurate input to the programme is needed. It is easy to use and to control, and may be especially suitable for children, who are likely to be familiar with it.

The trackball

Trackballs have been used in military and aviation environments, and are also used in some portable computers as an equivalent to the mouse. It incorporates a ball that is free to rotate in a fixed mounting. Receptors inside measure the rotation of the ball as it is moved by the user, and feed it into the system. Like the joystick the trackball does not need much space to work in, but users need time to learn the relationship between movements of the ball and the movement of the cursor on screen, as there is less immediate physical feedback.

The touchscreen

Touchscreens are an attractive option for point-of-sale and point-of-information CD-I discs, because they are easy to understand and use, and don't require a separate pointing device that can become detached from the player or damaged. In some touchscreens the screen area is covered with infra-red beams; when the screen is touched these are broken, and the system is able to recognize which spot on the screen is being indicated. Other systems consist of a touch-sensitive overlay on the screen. This may make the screen slightly harder to view.

Research has shown that users don't always point as accurately as they believe they do, and it's important to make sure that each hotspot is large enough and has plenty of space around it.

The keyboard

Keyboards are probably the easiest way of getting textual information into the system. But for many types of CD-I programme this would not be necessary, and you should not assume the presence of a keyboard. Many users find keyboards intimidating, or use them slowly by 'hunting and pecking'.

The graphics tablet

This is a touch-sensitive plastic board that generates on-screen images when a stylus 'draws' on it. If your programme requires the user to do more than select items – for example, to draw shapes or connect objects – it's worth considering.

Other devices

CD-I can support any input device that contains an X-Y pointer for moving the cursor about the screen, and a means of selecting items on the screen. New ways of doing this are being developed all the time. One exotic device that is likely to become more familiar is a specially wired glove that interprets the wearer's hand movements to point to and select items.

Physical controls such as buttons and the on/off switch form part of the interface. A slot may be provided to allow users to insert their personal smart cards. Such a card could be used as

- a 'key' to control facilities or data
- a record card for keeping personal data, training progress or game scores
- a security device
- an identity card – for, say, home banking or personalized training

Displaying interactions and events

A large part of screen design is purely graphic design, creating the right image for your programme and reinforcing its message. However, there are areas where you need to consider the interface first and graphic design second to ensure that your programme is as usable as possible.

Cursors

The cursor is the visual symbol that allows viewers to see what they are pointing at. This is a vital function in an interactive system; it's the precise point where the user meets the system in order to instruct it and make selections.

At its most basic, the cursor is a small arrow that moves around the screen when the user moves the control. As it moves over a hotspot the arrow changes colour or the hotspot is highlighted.

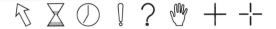

Figure 4.1 The cursor design can tell the user what the system is doing, or can give prompts for action.

Visible or invisible?

The CD-I system has special facilities to enable a cursor to be present at all times (see page 82). However, you may not want a cursor on screen all the time. It may be appropriate to turn it off during an audio-visual segment, to show the viewer that there are no opportunities for interaction until the segment has finished. The reappearance of a cursor would then be a sign that the user needed to choose what to do next.

Shape and design

A cursor need not be a simple arrow. By using a consistent set of changing cursors you can help the user to understand what kind of action the system expects. With some CD-I programmes the viewer has to make only a few simple selections; others have a more complex set of interactions. For example, in a programme that could accept text input the cursor could change to the I-beam shape used by word processors, showing that text should be typed in now. Or a cross-hair pointer, as used by computer graphics programs, could be used when graphics needed to be selected.

The cursor can also change colour, to show that it is over a hotspot or that it is now time to make a selection. Try to integrate the cursor into the overall design. It can represent any object that fits into the design – in a detective story, for instance, it might be a small magnifying glass.

Figures 4.1 and 4.2 show some suggestions for cursors that explain anticipated actions to users.

Confirmations

Some designers remove a redundant cursor from the screen once a selection has started running. In fact some have dispensed with

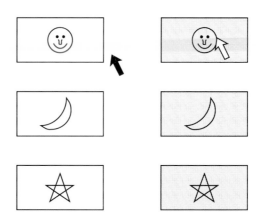

Figure 4.2 Menu items or buttons can change colour, showing the user that this is a hotspot.

the cursor completely and simply highlight the whole area being pointed at. If you decide on this technique make sure that you give users clear visual cues as to what they have chosen.

Others favour leaving the cursor on screen. If the situation is simply that no action can be made at that point on the screen, but there is an opportunity for interaction elsewhere, removing the cursor would be misleading to the viewer.

Viewers need to know that the CD-I system is responding to the selections they have made. Sometimes this will be obvious, because the response is immediate. However, in some situations it may take a perceptible time for the system to retrieve the information requested, and it may seem as if nothing is happening. Viewers' confidence in the system and their ability to control it can be shaken by pauses in which it is unclear what is happening.

Ideally the system should acknowledge all user input separately from performing the requested action. This acknowledgement should take the form not of a textual message but of a more immediate visible or audible signal. For example, when a shot is fired in a computer game, the player knows it has happened because of the sound of gunfire and the appearance of any damage to the target. A miss may be noted by a negative sound, a voice saying 'Oops!', or something similar.

Reassurance is important when the player is busy getting data or loading a display. It may help the user to display a small cursor that shows that something is happening – such as a clock. You can also reassure the user by designing the programme so that the cursor behaves consistently. When it gets to the edge of the screen, it shouldn't disappear – it should either stop, or reappear on the opposite side of the screen.

In the same way viewers of all types of programmes need to know that their selection has been registered and that they need not try again. The selected item could be highlighted and stay that way, or a special cursor could be used to show that the system is busy fetching the selection.

The viewer also needs to see whether the cursor is in the right place, or that now is an appropriate time to make a selection. The cursor could change colour or style to show that the selection is available and that all the viewer need do now is click.

Hotspots

A hotspot is an area on the screen programmed to respond to user input. When it passes over a hotspot, the cursor usually changes shape or colour to show the user where the hotspot is located.

In a quiz game the cursor might change from a question mark to an exclamation mark when the viewer finds a clue. This is achieved by defining two cursors in memory, then switching them over when the cursor moves onto the active spot.

Sometimes no cursor is on screen, but when the user moves the remote control down a menu, each option is highlighted in turn. This is because each of the menu selections has been predefined as a hotspot.

A hotspot can be any shape; in an animal picture it might be the shape of an elephant's trunk. It can also be any size, from the whole screen down to one pixel (the smallest picture element). Any number of hotspots can be used on screen, although only the one that the cursor is on will be active. Most hotspots are likely to be clearly defined 'buttons', labelled with the result that clicking on them would achieve – the video sequence that would be played, for example.

One rule of thumb is to have no more than six hotspots on screen, with each hotspot no smaller than a sixth of the screen. Trying to locate and activate a small hotspot using a remote control, when the player is at the other side of the living-room, can be difficult and frustrating. Selecting the wrong hotspot would add to the problem; it may be a good idea to have an inactive border between hotspots.

Hotspots need not be simple selection devices. Here are some options.

Sliders

Setting the volume for sound output, for example, might be achieved by an on-screen sliding control, moved either by pointing at it and directly sliding it with the pointing device, or by clicking many times to move the slider step by step. Sliders can be appropriate when the viewer needs to decide on a question of scale rather than make an on/off selection.

Invisible areas

Some hotspots will be invisible until they are selected. For example, a game might ask the child playing it to find where the cat is hiding, and offer text clues. The answer would become clear when the child selected the hotspot, which might be a piece of furniture.

Buttons

There are many different ways of designing buttons (Figure 4.3). As described above, clearly marked areas will be one form. Others may resemble checkboxes or radio buttons. Computer programs observe some useful conventions with these; if more than one item can be selected at once, checkboxes are used. If only one choice can be selected at once, radio buttons are used; any previous choice is automatically cancelled by the new selection. This approach is useful for such items as selecting the language for sound output.

Icons

Not all hotspots need to be labelled with text. In many cases a picture representing an object can be used instead. In computer terminology these are called icons.

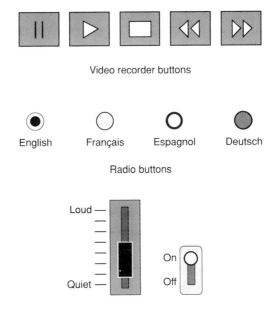

Figure 4.3 Some types of on-screen control icon.

Some people feel that icons are a clear, visual way of presenting information and options. Others feel that because the user has to translate them mentally into what they symbolize, they have the effect of distancing the user from the programme. This is almost certainly more true of games and simulations that engage the viewer's imagination than it is of more information-based CD-Is, such as training discs.

If you decide to use icons, try to choose ones that have an immediate and unmistakable meaning. Earlier in this century, the British road sign to warn drivers of a children's crossing was a torch, symbolizing knowledge. The driver was supposed to translate this as:

torch = knowledge = school = children = danger

Today, the sign has been changed to a picture of children crossing the road. This is a direct image that does not need translating.

There is a lesson here for designers of icons: don't be clever or allusive. Be straightforward. The point of using icons is to reduce the amount of information that users have to wade through to get to the information they want. The meaning of an icon should be immediately obvious to users who are not familiar with the system. For example, you can assume that most viewers are familiar with the symbols used on domestic VCR players to indicate play, stop, fast forward and rewind, and so you can use these to enable the viewer to control video segments or move through 'pages' of information.

Remember too, that the meaning of symbols varies from culture to culture. The sole of the foot, the left hand – these both have negative connotations in some societies. If you are preparing a programme for use in several countries, it would be wise to check all the icons you are using for cultural acceptability.

Designing a screen-based interface

This is perhaps the most discussed area of interface design. However, while there are many agreed findings, remember that much advice on interfaces comes from the computer industry. It assumes that the interface is for a computer, an ATM banking machine or a similar device. In CD-I, however, you should try to avoid designs that feel like a computer. For many people, computers have negative connotations. By presenting the audience with a computer-style interface, you may alienate them. If they see such an interface as part of a rolling demonstration in a shop, they may skip over your programme and choose something that looks friendlier or more fun. Your viewers are much more likely to be familiar with television programmes and the graphic style used in that medium.

As always, the most important considerations are the needs of your audience and the way you expect them to use the programme. For instance, guidelines on how to colour text are all very well, but if your CD-I is aimed at a home audience, sitting at least six feet away from the screen, is detailed text the best way to communicate with them?

Your design will also need to take note of the display capabilities of the CD-I system. These are covered in detail in the next chapter.

Using text on screen

Very few CD-I programmes will be entirely nonverbal. Words are needed to communicate with the viewer, both in the programme content and in the instructions on how to use the player and programme. This section is about the latter.

The words you use are going to be very important. Choosing the wrong word may make it much harder for viewers to understand what they should do to interact with the programme. However, too much detailed instructional text will clog the screen and end up making the programme even harder to use.

The best result will come from using a combination of simple, concise text and graphics to give the viewer a clear indication of what must be done and how any particular piece of information can be reached.

Menus

In many places you will provide menus, lists of actions that the system can perform at this point if the viewer chooses one of them. Ideally each item should be only one or two words, with a verb to say what will be done and a noun to say what it will be done to. Examples are 'Open file', 'Leave programme', 'Save data'. However sometimes you will only need noun phrases as the meaning will be obvious: 'Next page', 'Last page' and so on. You could also include a text message telling the viewer to make a choice.

Menus also allow the viewer to choose which section of the programme to go to. Although you may want to use words for this, pictures or icons should be used as well. To help the viewer know where he or she is in the programme structure your menus should have names that are used consistently throughout the

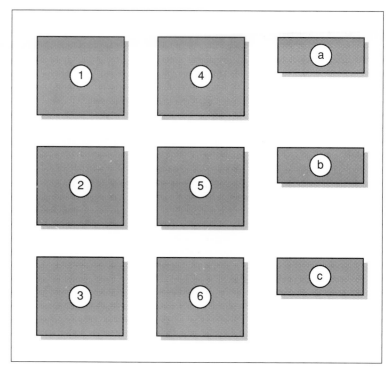

Figure 4.4 Equivalent menu items should have the same size and appearance on screen.

programme. For example, the opening menu could have the same name as the programme, and each choice listed in it could be the name of the next menu.

In programmes through which the viewer is allowed to wander at will, you should always include an option to go back one move, to the previous menu. This is an easy way of allowing the viewer to cancel mistaken choices, or to experiment with the programme without requiring an irrevocable commitment to choices that have been made.

The size and placement on screen of menu options is important. For items that appear on every menu, such as 'Quit the programme' or 'Go back to the beginning', find a suitable place and style and keep them there on every single menu. Equivalent items, such as different categories of animals, should be similar in size and shape (see Figure 4.4).

Disc navigation

Much interaction will be concerned with disc navigation, moving around the programme. What options are you going to give users at each place where the programme needs their input? Your programme design should show you where you want the user to be able to move – to the next section, to more detail on something in this section, out of the programme entirely, and so on. Options that you could offer include:

stop	restart
pause	repeat
jump to	go forward
go back	speed up
escape	slow down

Personalizing a programme

Interactivity is excellent for building up a highly personal relationship with otherwise impersonal data. For instance, a woman in a dental surgery might, by answering questions and entering details, build up a personal dental health profile. She can then use it to explore the kinds of treatment and advice appropriate to her age and dental history. An apparently different but actually quite similar use is in games, such as detection adventure games. Here the audience can explore hotspots on the screen. They can look for clues that lead down alternative paths, building up a picture of what has happened and eventually solving the case.

More complex interactions

At the next level, if it is appropriate, you may want to offer options that enable the user to:

delve deeper	move back up a level
cross-reference	return to a place marker
switch channel	

Badly thought-out interactions can be frustrating and irritating for the audience because they slow down the action, for little apparent result. But well designed interactivity is a motivating force, stimulating interest and personal involvement.

When you are building interactivity into the programme – first into the concept, then into the design – keep it simple to begin with. Simplicity and consistency are vital. You should consider whether every interaction and choice are necessary. The simpler the programme is, the easier it is to use and the more your viewers will get out of it. You should always design with the new user in mind, who has not played your disc before and may even be new to the medium – keen to explore it but frustrated by anything that is not easy to understand. Although it may seem tempting to add many ways of getting from one point on the disc to another, a single route is easier for the viewer to remember and use again.

Non-text-based interactions

The dialogue between player and viewer does not have to be text-based. Information can be requested by a voice-over, a cartoon character or a soap-opera heroine – perhaps in a song or a piece of graphics. Dialogue can take the form of:

graphics	animation
speech	audio
text	choosing
exploring a hotspot	drawing round an object
allowing time to pass	erasing an item
music	noises

Interactivity can give users control over aspects of the programme – to:

slow down	pick out
speed up	cut out
darken	synthesize
brighten	make louder
zoom in on	make softer
zoom out to	shape a sound
pan	shape an image
change angle or position	colour or re-colour
erase	'become' a character
join in	play along

Try to get the form of the dialogue to match the design concept you have chosen.

Using metaphors

Successful nonverbal interaction depends on the use of metaphor. In interactive design terms this is an integrated set of visual treatments that assist users in navigating the programme. A good metaphor makes it easier for the viewer to understand the underlying structure of the disc. For example, a CD-I about travel could use a page from an album of postage stamps to show that it covers many countries, and viewers could select a stamp to visit that country (see Figure 4.5).

Accompaniment

In some cases the viewer won't be responding directly to the system but accompanying it – for example, in a home karaoke CD-I disc or a keep-fit programme. Here the challenge is to ensure that sufficient interaction is built in to maintain the viewers' interest, but not so much that they are unable to concentrate on what they want to do because the programme keeps getting in the way.

Figure 4.5 Metaphors provide one way of communicating programme structure to the viewer.

Real time

A person sitting at the wheel of a driving simulator is 'driving' it in real time. The system will not slow down when things get tricky. It will present events at the speed at which they would occur in reality and the user will have to react equally rapidly. The picture changes constantly in response to the user's actions. Continuous real time is often used in games and simulations.

Screen design guidelines

Make it consistent

A consistent interface is learned and remembered more rapidly by users. Remember that consistency has to do with colour as much as with words (for example: 'If red was Help last time, why isn't it this time?').

Make it friendly

Whatever the programme, the best-designed interface will fail if it isn't friendly. The interface should invite the audience to explore the system. Don't force the user to remember chunks of information like 'I must press « to go back'. Put on-screen controls in 'buttons' if necessary. When users make a choice, give feedback – show them which item, colour, picture, etc. has been selected.

Make it obvious

Viewers need to be able to understand the programme structure quickly in order to be able to use it. Do not make menus over-complicated or use a metaphor for programme items that is difficult to understand or requires too great a leap of the imagination. If your programme metaphor seems strained and uncomfortable, do not use it.

Use colour carefully

Some programmes make heavy use of colour. Colour is, of course, a highly effective nonverbal method of giving information. (For example: 'Those areas are both yellow so they must be related'.) But be disciplined. If used randomly, colour can both distract and confuse.

Remember the demands of the screens on which your viewers are likely to see the programme. These may not be particularly sharp or well adjusted. The best approach is to experiment and test colour choices before finalizing the programme design. If your programme will be distributed internationally, experiment with both PAL and NTSC TV standards, as they may pose different problems.

Consider using different shades of the same colour, rather than outright colour contrasts. Ideally, screens that present text should have a background that gently recedes, so that the foreground material shows up as sharply focused. Blue is often used for screen backgrounds because its visual properties make it effective and fairly restful – grey and white are also used frequently.

Try the interface early on

Above all, try the interface out as early as possible, and invite comments from people who closely match your target audience in terms of experience, expectations and interests.

Test the interface in real conditions

Make sure you see the interface in the setting in which it will actually be used, such as the living-room. Light, distance, temperature and so on should all match the expected 'real' conditions of use. For example, if the user is likely to be wearing gloves while using an outdoor CD-I point-of-information system, your design will need to take into account the consequent reduction in accuracy of pointing and selection.

User errors

However well your programme is designed and implemented, users are bound to make mistakes and attempt to do things that can't be done. They may try to select inappropriate options, unintentionally press the wrong button, or do countless things you weren't expecting. Obviously an important part of the production of CD-I programmes is checking that they are sufficiently robust to stand up to this kind of abuse. But an equally important part of interface design is to minimize the number of mistakes by letting users know what they can't do and when they have done something wrong, and pointing them in the right direction.

It is important to do this, so that people can use the system without finding that they make a mistake and get stuck with no guidance about what to do next. It is even more important that any instructions do not add to the confusion and damage users' confidence in the system and their own ability to use and control it.

With most CD-I programmes, the best approach is to design it to do nothing until a selection has been made and to ensure that valid selections are clearly signposted. However, for some custom programmes with complex interactions, aimed at experienced users, it may be more appropriate to display a textual message to explain, for example, that the selected item was not available, or that the system cannot accept text at this point. Whatever approach you decide to take, you should at all times avoid the kind of cryptic error messages that make computers so daunting to use for many people.

Conclusion

Once a programme concept and treatment have been finalized, details of the design need to be considered. The interface between programme and viewer is an important area that needs attention at this point in the design process. You may already have clear ideas of things that you wish to attempt in interface design. Otherwise, you could look at other CD-I discs for inspiration.

The precise design of the interface is up to the developers of each CD-I programme. However, familiarity and consistency are an important part of ease of use, and developers should consider what has gone before. Companies that plan to produce many CD-I discs may prefer to create their own guidelines and stick to them in all the CD-I programmes they make.

5

Using video with CD-I

Introduction

A thorough grasp of the technical possibilities – and limitations – of CD-I is vital if programme designers are going to get the best use out of the system and their programme material. This chapter looks at the way in which CD-I systems display text, graphics and video, and the design implications of this.

It begins by describing the display capabilities of CD-I, a topic that encompasses both the physical aspects of the display and the system's ability to layer several images on top of each other in 'planes'. This is followed by detailed discussion of the ways in which different types of image can be used and stored by CD-I systems.

The CD-I screen

Screen resolution

The CD-I player's screen has the characteristics of a television screen. This has implications in terms of its resolution, the level of detail that it can show. There are several aspects to screen resolution that the designer must bear in mind.

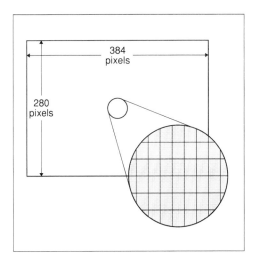

Figure 5.1 Resolution of a PAL-standard
screen. (A pixel is a picture element)

Of course, not all CD-I screens have the same resolution.
Although 'base-case' systems will have the normal resolution,
two further levels of detail can be used by programmes that will
be played on more highly specified systems. **Double resolution**
doubles the number of pixels (picture elements) on each line;
high resolution has twice as many pixels as normal resolution,
both vertically and horizontally. This has implications for the
size of the images you can use and the speed with which the
player can get them onto the screen.

Pixels

Resolution is largely, but not wholly, a function of the number of
pixels on the screen. The more pixels you have, the more detail
you can show on the screen (Figure 5.1). Computer screens have
a large number of dots relative to their size, giving a very high
resolution: 80 columns and 40 lines of text can be displayed
clearly and crisply on many monitors. There is very little flicker
because the computer monitor scans – replaces the image – at a
far higher rate than an ordinary television screen. The computer
screen is relatively small because it is designed for use at close
quarters; this too helps to keep the resolution high.

The more pixels there are on screen, the more memory is needed to store information about each pixel, such as its colour. Constantly refreshing those pixels from the computer's memory also poses speed problems. To use only 256 colours on a screen requires one byte to store each pixel's colour. (A byte is the standard unit of data, consisting of a sequence of eight, or sometimes seven, 'bits', or zeros and ones.) For a 360×240-pixel screen, this can add up to over 80 kilobytes. (A kilobyte is 1024 bytes.)

With a computer, this problem of balancing speed and economy of memory use on the one hand against resolution on the other is usually solved by limiting the number of colours that can be displayed. That way, the demands on the computer's memory are limited. The usual choice in a computer system is a large number of pixels (for crisp text and clear detailing) with a limited number of colours. This is appropriate for computer applications such as databases and spreadsheets, which mostly consist of text, numbers and diagrams.

Colour

Screen resolution depends on colour. Compared to a computer screen a television screen has a low spatial resolution. Yet it can display an almost infinite number of colours because it is an analogue medium – information is conveyed by continuously varying waveforms, rather than digitally. Television can therefore give the impression of high resolution because of the wide range of colours that is generally used.

However, on a television screen text is usually of low quality; the letters look ragged, because of the low numbers of pixels. Even high-quality video fonts can display only 40 or 50 characters across the screen.

Physical screen size

Designers also have to bear in mind how much bigger television screens are than computer screens. For this reason, people who are used to constructing graphical video screens will find CD-I easier to adapt to than those whose experience has been in designing for computers.

Designers whose experience is largely computer-based may also be used to thinking in terms of PC or Macintosh graphical environments. They need to adapt their thinking, or they will find themselves producing designs that prove extremely hard to implement on CD-I. For instance, windows within windows, each containing text, are a common feature of computer displays. But they rely on the ability of a PC to display text that is small yet sharp. On a CD-I display, using much bigger text, this kind of design is not nearly so effective because, relatively, there is less space.

The question of colour use shows again the marked difference between designing for CD-I and designing for computer displays. CD-I designers need to have a much firmer grasp of how the colour systems work.

Video formats and image compression

The selection of the appropriate image format has many important implications for CD-I design. Image format affects the final look of the screen, how quickly the image can be retrieved and displayed and how much space it takes up on the disc. Understanding these formats will help you to design better programmes. For example, you may be considering both animation and full-motion video (see page 77) for a programme segment. Your budget could allow either and both would work well to illustrate the point your programme needs to make, but when you consider how much disc space each would require, you could find your decision made for you.

As discussed above, the CD-I player's display has the characteristics of a television screen. Once the design team has compiled the list of production assets (visuals, audio, and so on),

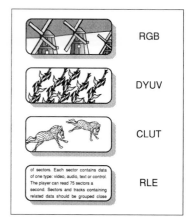

Figure 5.2 Types of image compression.

it becomes necessary to decide how they are going to be stored. This is a fundamental design issue.

A screen usually consists of up to 107,520 pixels (on a 384 × 280 pixel screen). A high-resolution screen contains four times as many pixels (twice as many horizontally and twice as many vertically). A laborious way of digitizing a picture would be to record colour and brightness data for each pixel in turn. Compression of images short-cuts this process. This usually works by recording the differences between adjacent pixels in each row rather than their absolute values, and by reducing the number of colours available in the palette.

Pictures can be digitally encoded in a number of ways (Figure 5.2). Most of them result in some compression of the amount of data required to store an image or sequence of images. The discussion below shows that if the system chosen is DYUV, then the colour will be high-quality and natural-looking. But the algorithm that compresses the image to a manageable file size will also blur the edges of objects in the image. On the other hand, if a wide colour spectrum is less important than sharp definition, a CLUT image is the obvious choice.

High-quality natural images – RGB

Highly detailed still images – for example paintings – might be encoded using the RGB 5:5:5 (Red Green Blue) method. RGB 5:5:5 gives a range of 32,768 colours. It simply records the colour

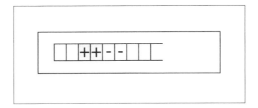

Figure 5.3 Image compression: DYUV. The system does not record an absolute value for the colour and brightness of each pixel, but only differences, if any, between these quantities and those of the preceding and following pixels.

data for each pixel, and therefore achieves no compression. Using the RGB format limits the number of visual effects at your disposal. You should use it only when an image of the highest quality is essential. An RGB image takes up around 200 kilobytes or 100 CD-I disc sectors.

Natural images – DYUV

Natural images (such as fields of waving corn) are compressed using Delta Luminance Colour Difference, known as DYUV. DYUV compression exploits the fact that the data for neighbouring pixels are often similar, so instead of recording each pixel separately it records only the differences between pixels, line by line (Figure 5.3). It also exploits the fact that the eye is more sensitive to differences in brightness (luminance) than in colour (chrominance). DYUV saves data space by recording less information about colour changes than about brightness changes. A DYUV image takes up around 90 kilobytes, or 45 CD-I disc sectors, depending upon the size of the screen. The type of image is irrelevant to its final compressed size.

Graphic images – CLUT

Graphic and computer-generated images can be compressed using Colour Look Up Tables, or CLUTs. These are tables of variable size containing a selected palette of colours (Figure 5.4).

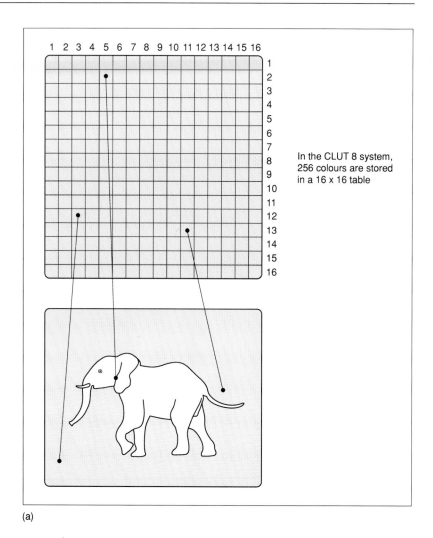

In the CLUT 8 system, 256 colours are stored in a 16 x 16 table

(a)

Figure 5.4 Image compression: Colour Look Up Table (CLUT). (a) When an image is displayed, the colour of each pixel is not given an absolute value. Instead its position in the colour table is specified. (b) Dynamic updating. In the time it takes the dot that 'paints' the screen to go from the end of one line to the start of the next, a new CLUT can be loaded. Changing CLUTs permits a wider range of colours to be employed. Also, a given pixel can be assigned a different colour without changing its CLUT number, which is useful for some animation effects. Thus the colours of this snake can shimmer as the CLUT tables are alternated.

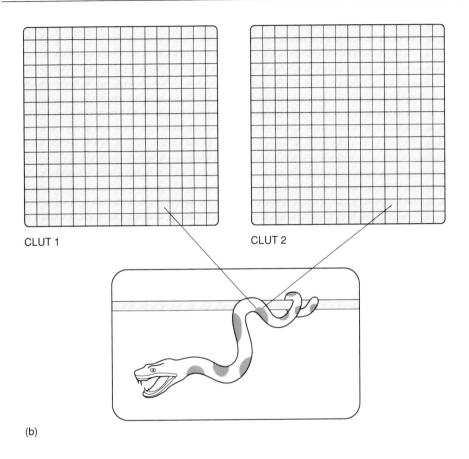

CLUT 1

CLUT 2

(b)

CLUT 1 sky tones

CLUT 2 sand tones

(c)

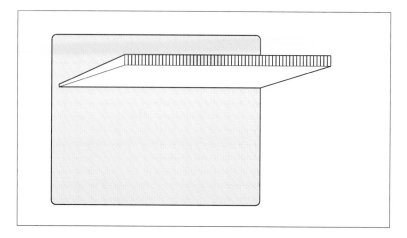

Figure 5.5 Image compression: Run Length Encoding (RLE). Instead of recording a value for each pixel, this system records the number of pixels having the same colour value as the given pixel.

Each colour has a unique position in the table, and colours in an image are recorded using their address in the table rather than their absolute RGB colour value.

A larger table allows more colours in an image but the result takes up more space. CD-I supports three levels of CLUT image: CLUT 8, CLUT 7 and CLUT 4. The number indicates the number of bits of data used to store the CLUT address of the colour for a pixel. A CLUT 8 image uses 256 colours chosen from a palette of over 16 million. Additional flexibility is provided by updating the colours in the table on a line-by-line basis, giving a choice of over 1000 colours per screen. CLUT 8 and CLUT 7 images take up approximately 90 kilobytes, or 45 CD-I disc sectors, again depending on the size of the screen.

Cartoon images – RLE

Cartoons and animated sequences are compressed using a technique known as Run Length Encoding, which further reduces the size of CLUT images. It records the colour value of a particular pixel and then says how many of the following pixels are to be of that colour (Figure 5.5). It is less subtle in its effects than DYUV, but ideal for animation, where cartoon images may have relatively large, flat areas of colour.

An RLE image, depending on content, may typically take up approximately 10 kilobytes or five sectors, although this will vary with the amount of change and contrast in the image. The speed at which an animation plays is closely related to its size and colour complexity (that is, the number of colour changes on each line of the image). Each colour change takes up two bytes.

Higher resolutions – QHY

A further compression type is available for use in extended players, to display natural images at an apparently higher resolution. QHY (Quantized High-resolution Y) images can be used only on players with this extra hardware. QHY works by enhancing the luminance or Y signal of the image and interpolating it between the chrominance and luminance values already stored as a DYUV image. Where luminance differs greatly between two DYUV pixel values, QHY interpolates a correct value to achieve the effect of a sharper image.

Combining compression types

The total area of the screen can be divided into a number of horizontal bands known as subscreens, with a different compression or coding system operating in each. So you can use the most suitable technique for each visual resource, making the most of your images and minimizing the data space required (Figure 5.6).

CLUT graphics palettes

Getting the best out of CLUT graphics palettes requires careful management. When a natural image, such as a photograph, is scanned into a file the scanner can grab a very large number of

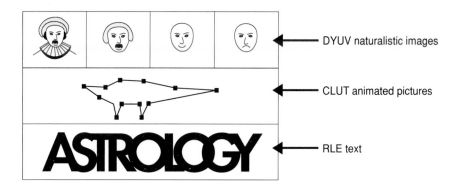

DYUV naturalistic images

CLUT animated pictures

RLE text

Figure 5.6 Using different compression types in one image. Each picture plane can be divided into horizontal bands, called subscreens, using different compression types.

colours; up to 16.7 million different colours can be stored if 24 bits are used for each pixel. CD-I software contains an algorithm to reduce the number of colours in an image to CLUT 8 level, where 256 colours, requiring eight bits each for storage, are chosen by the system, as the colours that best match the colours in the original image.

You do not have to leave this process to the computer system. You can start with black and white and create every colour in the palette, using RGB colours to draw with, thus creating the palette as you create the image.

Problems may occur when you have loaded in a CLUT 8 256-colour palette, but now want to add colours. You have to recompress the image, losing quality as you do so. Furthermore, you may now be using a different palette, and the appearance of the two may clash on screen.

Exploiting the limitations and opportunities of the way computers store graphics is a challenge to which artists have risen. An image originally created in 100 shades of brown can quickly be displayed in 100 shades of purple or orange. This technique is often used for special effects and for animation.

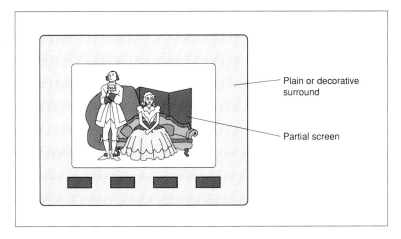

Plain or decorative
surround

Partial screen

Figure 5.7 Partial-screen motion. A partial-screen video picture uses proportionately less memory than the whole screen would, and can be just as effective.

Compressing motion video

The CD-I player has a special chip for processing moving video data, and half a megabyte of memory for storing and handling moving pictures, in addition to the one megabyte reserved for other types of video. (One megabyte, symbol 1MB, is approximately one million bytes.) However, moving video is fairly expensive in disc space, and it is worth considering two ways in which you can reduce the amount of memory it uses, both on the disc and in the player.

Partial-screen video

Using partial-screen video can make big economies. If you use 40% of the screen area for full-motion video, you also need only 40% of the memory you would have required for the full screen (Figure 5.7). So, for example, you could use motion in a window on a screen that also contains some descriptive text (see the section below on picture planes, page 80). Partial-screen mo-

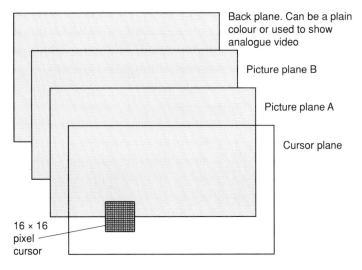

A and B are used for most
purposes–pictures, graphics,
text, menus etc. Each uses up to
half a megabyte of the player's memory.

Back plane. Can be a plain
colour or used to show
analogue video

Picture plane B

Picture plane A

Cursor plane

16 × 16
pixel
cursor

Figure 5.8 Picture planes.

tion video does not have to use the special motion video chip
and memory; it can instead use the one-megabyte player
memory normally used for other types of image.

Partial-motion video

Partial-motion video uses a rate of less than approximately 20
frames a second. At this speed, motion starts to become slightly
jerky. This sort of motion is perhaps best described as resembling a series of rapid cuts between still pictures. It may be a
useful technique – for example, in demonstrating a procedure
for replacing the gasket in an engine: the slow, slightly jerky
motion is an effective way to show the audience how it is done.
Like partial-screen motion, partial-motion video does not have
to use the special motion video chip and memory, but can use
the one-megabyte memory of the player that is normally used
for other types of image.

CD-I in the home.
'Watching with Mother' – with a difference!
Now TV can be active fun rather than passive
adding creativity to learning.

A typical painting sequence from 'Cartoon Jukebox'. The user selects a colour by dipping the 'paintbrush' into the 'palette' and then paints the chosen area of the picture.

The cartoon then runs along with the song with the colours as chosen by the player.

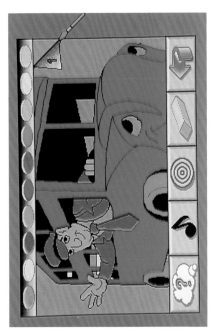

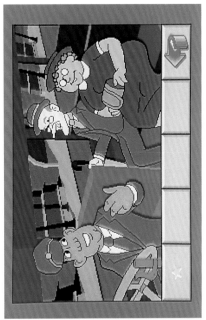

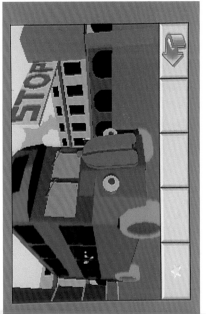

Learning about 35mm photography via CD-I with Time-Life Photography workshops.

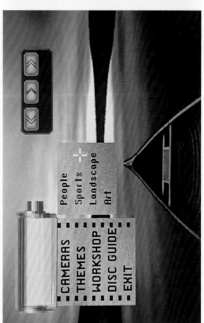

How to set the 'exposure' and then take a 'shot' to check that it is correctly 'exposed' before using your own camera! This means that every photo is right first time!

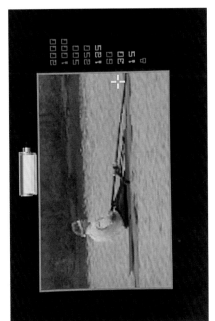

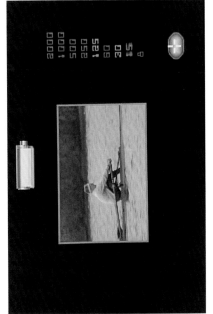

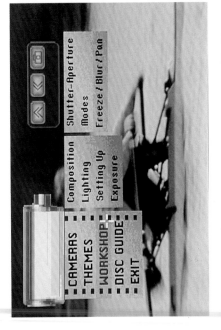

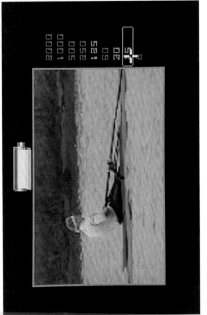

A typical surrogate travel CD-I application with appeal for art lovers — The Renaissance of Florence.

Visiting Florence to see samples of Michelangelo's work in situ.

Golf for the enthusiasts!
Play the courses of Palm Springs, Florida
from the comfort of your armchair –
it's almost as good as being there!

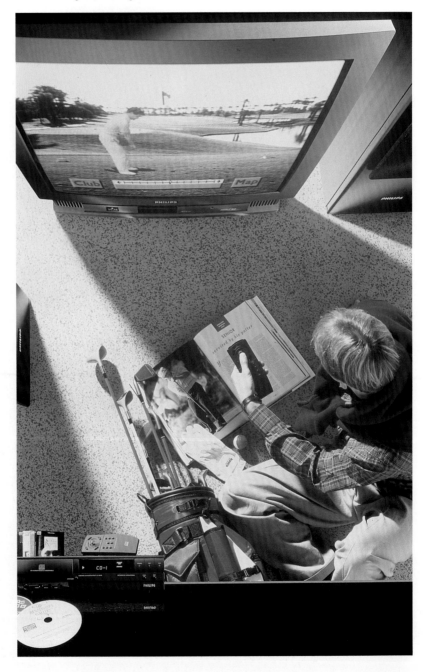

Video planes

Video planes are an important part of the design vocabulary of CD-I because of their creative potential. They carry over from the idea of multimedia as a slide show with accompanying sounds. They are like the slides stacked up in the carousel of a projector (Figure 5.8). However, with CD-I the relationship between one 'slide' and the next has become dynamic. The transition from one plane to the next can be made in various ways (fades, wipes and so on). In addition, parts of up to three planes can be seen at the same time by use of mattes and transparencies.

The planes can be used to hold moving images, an analogue video backdrop and the cursor, so the comparison with a slide carousel is left behind, and the planes become a dynamic multimedia design feature. The relationship between the various types of images displayed on the planes can be creatively exploited in all sorts of new ways.

Planes are especially flexible because they are not fixed. They can be swapped, reordered and so on. Remember too that each image plane can also be divided into horizontal subscreens that can be differently coded – so adding a further level of flexibility. Planes can be used in many different kinds of ways, such as:

- overlaying one image on another, as in montage, to create combinations of images and move one object over another
- creating visual effects such as wipes, dissolves and fades
- highlighting, as in menu choices
- loading and hiding labels, then remapping them and revealing them when they are needed
- windowing from a CLUT image to a DYUV image, thus achieving a mix of CLUT and DYUV images on screen together
- building up composite pictures

In using the picture planes to manipulate video images, you can imagine them stacked behind each other. Of the four planes available, three are used for images (Figure 5.9).

This partial-motion image uses two picture planes.

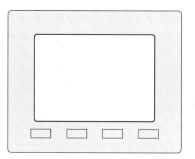

Plane A contains a flat surround and menu buttons. A matte in the centre acts as a hole through which you can see.......

.... a moving partial-screen image on Plane B.

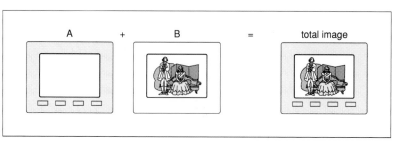

Figure 5.9 Use of picture planes.

The plane at the back is used to display full-motion video pictures. If you were simply using CD-I to show movies, this is the plane you would use. It can also be used as a background, given a colour, or used to play video from outside sources (that is, not from disc or memory).

The next two planes are used for other types of image. Each plane holds up to 512 kilobytes of data. You do not have to use both, but they provide many interesting and subtle techniques, including full- and partial-motion effects.

The front plane is the cursor plane. This is like an invisible pane of glass across which the cursor moves. The cursor can be any shape that can be contained in a 16 × 16 block of pixels.

You can manipulate the four planes in a number of ways to produce interesting and imaginative visual effects. These methods can be divided into single-plane effects (Figures 5.10–5.14) and two-plane effects (Figures 5.15–5.17). Two-plane effects can also be used on three planes. The effects dicussed here are the standard ones described in the Green Book: you can invent others as required.

Single-plane effects

Cuts

Cut from one image to the next. Rapid cuts from one frame to the next give the effect of full motion. The normal frame rate required to achieve smooth motion is between 25 and 30 frames a second.

Partial updates

Cuts involving only a part of the screen.

Scrolling

Involves the screen area moving across a larger image in a horizontal or vertical direction. Imagine the screen area as a square hole through which you can see only part of the image you want to present. Scrolling is the process of moving that square so that all of the image is eventually seen.

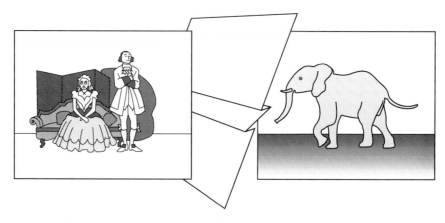

Figure 5.10 One-plane cut. The cut is not quite instantaneous.

Mosaic effects

These involve manipulating the pixels of the image to make it disintegrate into blocks or to magnify and coarsen it.

Pixel hold: Retains the whole picture but makes it look granulated by taking the colour and tone of one pixel and extending it to neighbouring ones. This process can be repeated until the whole picture seems to break up into a number of coloured squares.

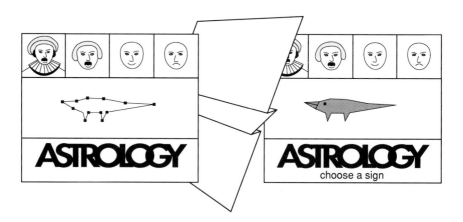

Figure 5.11 Partial update. Only one subscreen of the picture plane is updated.

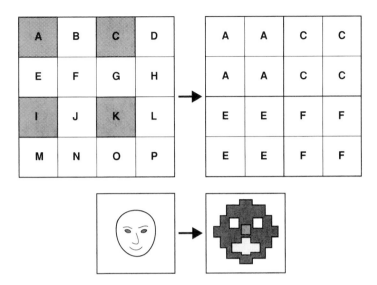

Figure 5.12 Pixel hold. The image is kept the same size while the definition is coarsened. This can be repeated to dissolve the image into large coloured squares.

Pixel repeat: Magnifies a portion of the total screen image without providing greater detail. The result is a larger but coarser image.

Fade up or down

Increases or decreases the brightness of an image between black and full intensity, through 64 levels. Different areas of the total screen image can be given different levels of brightness, so that only parts of it fade up or fade down. This effect is more often used with two planes.

Two-plane effects

Note that you cannot use two planes if you have used the RGB method to encode images, because an RGB picture uses all the memory for both planes.

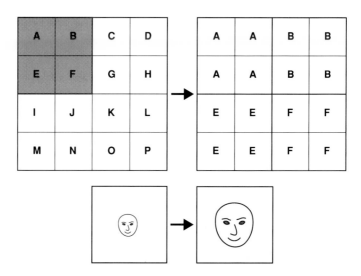

Figure 5.13 Pixel repeat. The image is enlarged, with a consequent coarsening.

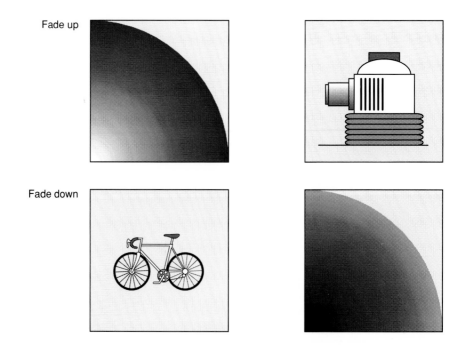

Figure 5.14 Fade up and fade down. Fade effects can be fast of slow, as desired.

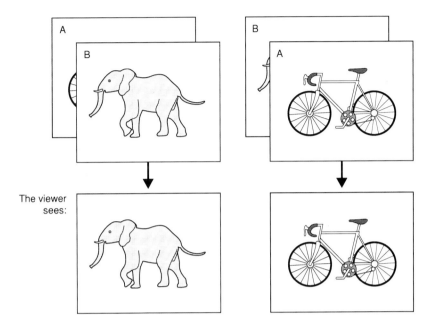

The viewer sees:

Figure 5.15 Two-plane cut. Picture planes A and B can be imagined to be instantaneously transposed so that the scene visible in the foreground changes.

Fade

Fade, as described above for one-plane effects, can also be used with images on two planes.

Cut

This can be used in the same way as for a single plane, described above. In addition, it is possible to cut from one plane to the other. If both images are held in memory, the cut from one to the next is instantaneous.

Wipe

The image on the front plane is progressively replaced by the image on the back plane, following a straight edge that can proceed from any side of the picture to the opposite one (for example: top to bottom, left to right and vice versa). There are several species of wipe:

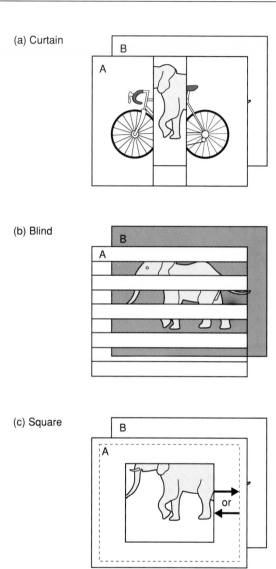

(a) Curtain

(b) Blind

(c) Square

Figure 5.16 Wipes progressively replace the image on one plane by the image on the other. (a) Curtains. The front screen can 'open' or 'close'. (b) Blinds. More amd more of plane B can be seen as the 'slats' of plane A disappear. (c) Square. The square 'hole' in plane A can shrink or expand.

Curtain: a wipe that opens out the image on the front plane, like a pair of curtains opening, to reveal the image on the back plane.

Blind: a wipe in which the front plane is like a Venetian blind, which opens to reveal progressively more of the image on the back plane as the slats disappear.

Square: a square hole of progressively increasing size appears anywhere on the front plane. (This is a wipe from the centre of the square outwards to all four sides simultaneously.)

Transparency

Transparent areas allow the back plane to be seen through the front one. For example, the front plane might be occupied by the menu screen, in which a transparent window allows you to look through to a moving image on the back plane. Two types of transparency are available to the designer:

Chroma key: This is a method of making a selected colour or colours on one plane transparent. This is a well-known technique in the video industry, and can be used to isolate an image on the front plane so that it can appear against whatever image is held on the back plane.

Mattes: A matte is an area of any shape in which the pixels of the image are made transparent. Mattes are used to achieve dissolves in which the image on the front plane melts uniformly into the image on the back plane. You can also use mattes to make a window in the front plane through which the image on the back plane can appear.

Using text with CD-I

Text is held on the disc either as characters or as graphics, which are transferred to the screen in the same way as other graphics. In most cases, an RLE image is actually better for achieving clear definition than a CLUT or a DYUV image.

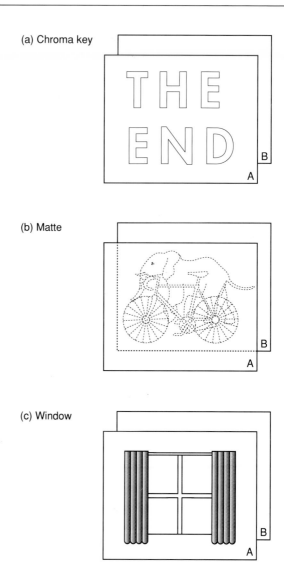

(a) Chroma key

(b) Matte

(c) Window

Figure 5.17 Transparency effects. (a) Chroma key. The letters on plane A were originally green. Green pixels are selectively made transparent to show some of plane B. (b) Matte. The pixels of the image on the front plane become transparent to achieve a dissolve from the front to the back image. (c) Window. The selected window area on the front plane is made transparent to show some of the moving image on the back plane.

ASCII text

Text can be stored in ASCII format, a universal computer format for storing text in the plainest form. When text is stored as ASCII characters on the CD-I disc, the font resident in the microprocessor of the CD-I player is used to display the characters on screen. This is the font used to display system messages on screen, and is suitable for similar uses within other applications.

Using fonts

You may want to use a typeface other than the basic system font used to display ASCII text and system messages. Other fonts can be downloaded at runtime, but memory in the player must be taken into consideration – a complete font, including Roman, italic and bold characters, can take up a lot of space, and a new point size counts as a new font.

Anti-aliasing text

The relatively low resolution of television screens means that the small details of letters don't come across very well – edges become jagged. The technique of anti-aliasing solves this problem. It uses a gradation of colour to fool the eye into seeing a smoother edge to the character. For instance, with black characters on a white screen, the four pixels next to each black pixel can be coded as shades of grey (Figure 5.18). Instead of an intense black/white contrast between character and background, this gives a gradual blending that makes the text look smoother. In effect, it produces a kind of dropped shadow. Paintbox graphics systems can carry out this process automatically, producing all-colour or all-luminance anti-aliasing, or a combination of both.

The CD-I designer buys the anti-aliased font and calls it at runtime (anti-aliasing is part of the fonts design rather than a runtime function). Remember that an anti-aliased font takes up slightly more space than an ordinary font and takes a little longer to load into memory.

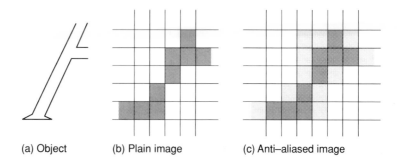

(a) Object   (b) Plain image   (c) Anti–aliased image

Figure 5.18   The 'staircase' effect produced on diagonal lines and edges by the limited resolution of the screen is combated by anti-aliasing. (a) The original object has smooth edges. (b) A 'staircase' effect is produced when it is represented by an image composed of pixels. (c) Neighbouring pixels are automatically shaded in intermediate tones, reducing the 'staircase' impression.

Creating video images

When CD-I designers talk about video they can mean several things, including:

- still images such as slides, which will be digitized
- still frames of video tape
- partial-motion video
- full-motion video

It is important for potential CD-I designers to appreciate the different kinds of production path that can be used for video images.

Let's take the example of a small production company that until now has been producing programmes on videotape for corporate customers. Usually the company hires in expertise or equipment that it doesn't have. A customer asks them to produce a point-of-sale CD-I, consisting of an interactive slide presentation.

The company has access to a Quantel Paintbox system and a designer, at reasonable cost. The staff are used to television-based production methods, so they decide to use those as far as possible. They use Quantel to process the images they want, adding graphics as appropriate. Then they dump the Quantel graphics to VHS videotape and do off-line editing in the stand ard way. Next, they show the client a preliminary 'slide show' – the off-line edit that is still on VHS. He asks for a few changes and they go back to their production list to trace the original Quantel graphics and change them. (The role of the production list is dealt with in more detail later – see page 151.) Once the client has approved the off-line edit, again on VHS, the company transfers the work to a CD-I Starter Kit and starts development in the CD-I environment.

This case study illustrates the way in which people can very largely choose the CD-I production path that suits their client, their resources, their experience and their designers. It's not necessary to radically change working methods that are familiar and efficient. These ways of working can be adapted so that they coexist with new CD-I development methods. What was, of course, important in this case was that the client should be fully briefed at each stage about what the presentation would and would not show.

An important aspect of this method is that it allowed all concerned to pass around a VHS videotape containing the work in progress. They could take it home or to meetings and show it to people without having to have CD-I equipment constantly available. It was a practical and successful way of achieving the results they wanted.

Mixing different kinds of video

Many designers will want to mix full-screen moving video with other effects from the CD-I multimedia palette. For instance, since the video images can appear in any rectangular area on screen, designers can run video with captions or other information on the same screen. Since the fidelity level of the video, and how much screen space it takes up, can also be adjusted by the designer, higher levels of audio or other program data can be accommodated in the datastream. This will form part of the tradeoff process, discussed later in this handbook (see page 117).

Full-motion video appears in an image plane behind the two standard image planes in a CD-I system. The designer has control of the windowing from one of these planes to the other, so FMV can be incorporated seamlessly into a CD-I programme.

Full-motion video on CD-I works differently from video on, for example, a LaserVision system. A full frame of video is encoded – this is often called an 'absolute' frame. For the following 'differential' frames, only the differences from the absolute frame are stored. Then the next absolute frame is stored. Whether you need a few absolute frames and many differential ones, or many absolute frames and a few differential ones, depends on how much and how fast the video picture changes.

Because of this storage method, video sequences cannot normally be rewound or played backwards. For certain operations you will need to specify the nearest full or absolute frame of video information rather than just the nearest frame.

It is essential to capture both video and audio material at the highest level possible. You can reduce the resolution later, if it proves necessary, when storing the material on the disc.

Scripting, shooting and editing video are often the most expensive processes associated with a multimedia production. Because CD-I is such a flexible medium, pieces of video can be reused in different CD-I packages with different combinations of text, graphics and audio. This can greatly reduce the cost of developing a series of CD-I discs for different customers.

Pace

When you use the disc to load video images and to change from one to another you use more memory. However, the pace of a CD-I programme owes more to the frequency with which the screen changes than to the size of the screen area changed. It may be more effective to make frequent, small changes to the screen presentation than infrequent, large ones. For instance, a large number of small pictures can be loaded onto the screen, hidden, remapped and revealed when they are needed.

Video design

When designing video for CD-I remember that, because it is an interactive medium, it can be treated differently in design terms from a conventional TV or computer screen.

For instance, using windowing techniques users can 'pan' across a visual field far wider than the actual screen. This larger picture might be, for example, a diorama of London. But unlike the kind of pan they will be familiar with from film and television, this one will be under their control. The screen becomes a viewing window that they can move up, down, right or left to see what the larger picture consists of.

Note also that certain video effects, such as zooming or scaling pictures up and down, can be achieved very quickly by simply resizing the screen pixels. However, you need to be aware of the effect this may have on the quality of the image. Zooming and scaling do not take place in the player – they are effects achieved during authoring.

How much video?

A key factor in design for CD-I is an appreciation of the true impact that choices such as the size of windows have. A prime concern is the proportion of the player's bandwidth – the amount of data that can be moved through the player at any one time – that is used. This can be less than might at first seem to be the case. For example, if half the screen height and half the screen width are used for moving video, this takes up 25% of the screen area – not 50% – leaving more bandwidth for other aspects of the programme.

Video link screens

If the programme needs to access the furthest part of the disc and load a complete new screen, it will spend a couple of seconds doing so. This time can be disguised by the use of link screens, which can, for instance, count down or use some other kind of animation technique to exploit the delay, in order to increase audience anticipation.

Designing for PAL/NTSC compatibility

The different national television standards have always been an area of difficulty for television and video producers and designers. CD-I aims to eliminate some of the problems by allowing discs to be playable on both the major standards, NTSC and PAL. In theory, CD-I discs can be played on any player anywhere in the world, but certain qualifications of this statement have to be made. The television standards are:

- PAL (Phase Alternating Line), used in Great Britain, most of Europe, Africa, Australia, China and South America: 625 lines, 25 frames a second.

- NTSC (National Television Standards Committee), used in the United States and Japan: 525 lines, 30 frames a second.

- SECAM (*Système Electronique Couleur Avec Mémoire*) used in France, the USSR and Eastern Europe: 625 lines, 25 frames a second.

You can design discs to work on either PAL or NTSC, or in compatibility mode, which works on both.

When you work in compatibility mode, you should remember that the screen size for a PAL set is 384 × 280 pixels, while for NTSC sets it is 360 × 240 pixels.

In terms of memory, one DYUV PAL image takes up about 100 kilobytes, while one frame of an NTSC image takes up 80 kilobytes.

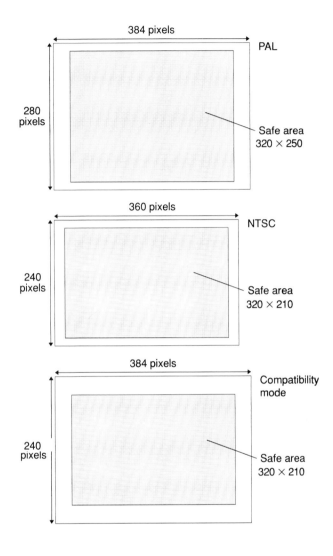

Figure 5.19 PAL, NTSC and compatibility modes, with their corresponding safe areas.

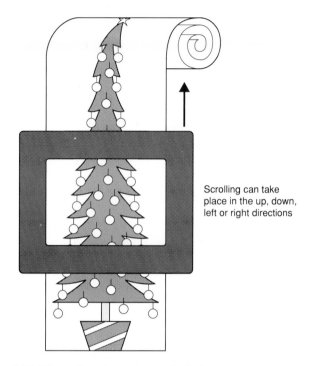

Scrolling can take place in the up, down, left or right directions

Figure 5.20 When the picture is scrolled, the screen acts as a window onto a larger image that is moved past it.

This means that, if you show a CD-I image designed for a PAL set on an NTSC set, you cannot see the outermost parts of the image. Add to this the fact that with pictures of either type it is anyway not advisable to use a certain area around the edges for important purposes. You should keep important text or menu buttons inside what is called the safe area (Figure 5.19).

Table 5.1 Screen sizes and safe areas for different television standards.

Signal/mode	Screen size	Safe area	Country
PAL	384 × 280	320 × 250	Europe (except France), Australia, Africa, China South America
NTSC	360 × 240	320 × 210	USA, Japan, Far East
Compatibility mode	384 × 280	320 × 210	All above

In practice, you should use the signal type that is appropriate for the market at which your title is aimed. If the title is aimed at markets that include both PAL and NTSC signal types, you should use compatibility mode.

Table 5.1 shows the safe area for each signal type. Obviously, if you are designing in compatibility mode you make the screen take up the whole of the area used by PAL pictures, which have more pixels, but only use the area that would be safe in NTSC pictures.

Scrolling

There are two kinds of scrolling (Figure 5.20). At this point, we will not enter into a detailed explanation of both – that is given in *The CD-I Production Handbook*, the next volume in this series. Designers simply need to know at this stage what the difference between the types of scrolling is. The chief difference is that dynamic scrolling involves a further degree of data-handling, in that it also uses the disc.

In static scrolling, the material that is not on screen is loaded into the player's memory. When the user triggers the remote control, the next portion of the visual is loaded and displayed. The redundant portion is stored in memory.

In dynamic scrolling, there is far more material to be dealt with. The player's memory is used, but the rest of the material has to be stored on disc. A relay system operates. When the user uses the remote control, visual material is loaded from memory to the screen. The spare memory freed by the material that has moved to the screen is filled up by material from the disc.

Because of the level of disc-searching it generates, dynamic scrolling leaves very little room for processing audio from the disc. For this reason, audio has to be stored in memory. One way round this limitation is to use audio samples (see page 104) to accompany dynamic scrolling.

Conclusion

CD-I includes enough ways of handling visual images to provide a means of producing almost any design. Understanding all the formats and their usage provides CD-I designers with an important part of their vocabulary and enables realistic plans to be made for the production of programmes that exploit the medium to the full.

6

Using audio with CD-I

Introduction

As you might expect, CD-I, a medium developed from the highly successful CD-DA digital audio standard, can handle sound particularly well. But 'CD-quality' sound is by no means the only standard to which CD-I can reproduce sound. Several levels of sound quality are available, allowing the designer to choose the one that best achieves the effect desired, while leaving enough space on the disc and enough bandwidth for the other parts of the disc, such as the images.

Sound in the CD-I system

The audio system contains two separate decoders for reading the data from the disc and converting it into sound coming out of the player. The **CD-DA decoder** makes it possible to play ordinary audio compact discs. The other, the **ADPCM** (Adaptive Delta Pulse Code Modulator), deals with digital sound data produced specifically for CD-I.

Audio material for use on CD-I discs, whether it originated as an analogue or a digital recording, must be converted to ADPCM format.

Sound quality levels

You can use any of four levels of audio quality on CD-I discs; these formats are discussed in more detail below. Once again, the greater the fidelity to the original that you require, the more data has to be stored, and the more disc space is used. For example, if you wanted to play *Così Fan Tutte*, you would use a high sound quality, whereas if you wanted to play speech, you would use a less space-consuming sound quality.

The use of an excessively high sound quality can prove to be a waste of disc space and bandwidth. How many of the players used for your programme will be hooked up to audio equipment sufficiently good to make the use of high-quality sound worthwhile?

The subject of your programme should give you some ideas about what type of sound is appropriate. An opera programme is more likely to be played through a quality hi-fi system and to need the best sound available, whereas a children's title may not be played on the highest-quality equipment and may not have such discriminating users.

Sound channels

Another aspect of sound quality is the effect it has on the number of channels you can use. If you use a lower-quality sound format, you may be able to squeeze in up to 16 channels. This means (for example) that you might have parallel commentaries on a travelogue in a dozen languages.

Stereo sound will use up twice as many channels as mono. The more sound channels you use, the fewer other things you will be able to have going on at the same time on your disc.

Sound formats

CD-DA

For the highest-quality sound, CD-DA sound quality can be used. This is the quality you get on CD audio discs, and is about as flawless as it is possible to get.

You would be unlikely to require such high-quality sound for a normal CD-I production, although if you were making a CD Ready disc – an intermediate format that is essentially a sound recording plus some electronic sleeve notes – you might use it.

You can get a little over one hour of sound of CD-DA quality on a CD-I disc, but as it would use up all the bandwidth, nothing else would be able to happen.

Table 6.1 Characteristics of CD-I sound formats.

Level	Data rate	Channels	Fraction of CD-I data rate used	Time available per channel
CD-DA	20 kHz	1 stereo	1/1	1.2 hours
CD-I ADPCM				
Level A	17 kHz	2 stereo	1/2	2.4 hours
		4 mono	1/4	4.8 hours
Level B	17 kHz	4 stereo	1/4	4.8 hours
		8 mono	1/8	9.6 hours
Level C	8.5 kHz	8 stereo	1/8	9.6 hours
		16 mono	1/16	19.2 hours

Level A

For slightly less high-fidelity sound, use Level A. This is equivalent to the sort of sound quality you get when you play a new vinyl LP on good-quality equipment, although without any of the background hiss. Level A stereo takes up 50% of the datastream, leaving the other half for video, graphics, text and so on. Level A mono takes up only 25 percent. You can get two hours of stereo or four hours of mono Level A sound on a CD-I disc.

Level B

Level B sound is equivalent to a first-class FM radio broadcast, and it takes up 25% of the bandwidth. You can get up to four hours of stereo Level B sound on a CD-I disc.

Level C

Level C is more like AM radio received under good conditions, and is adequate for speech. In its mono form it occupies only 6% of the bandwidth, leaving 94% for other things. You can get up to 19 hours of Level C mono sound on a CD-I disc.

Soundmaps

Soundmaps are short sound sequences (such as a beep used to confirm when an item has been selected from a menu) that are stored in the memory of the player and not on the disc. This means that they can be accessed instantly without the player having to go back and read the data from the disc. You can increase the playing time of a soundmap by making it loop, to achieve (for example) prolonged applause. Soundmaps may be stereo or mono and in any sound format except CD-DA.

Using audio samples

When it comes to working with audio on CD-I, there are ways of making very efficient use of the CD-I's memory. One of these is called sampling, because it uses short snatches of music called samples. These are sometimes specially written for use in the CD-I programme, and sometimes taken from earlier recordings; you will need to take care of copyright issues here. Note that sampling in this sense, although using the same technology and technique as the sampling process by which CD-DA recordings convert sound into digital information and back, is quantitatively different, in that it involves sound objects that may be anything up to several seconds long.

Each sample is an individual file, taking up about 150 kilobytes of memory and comprising a self-contained segment of sounds, perhaps a few bars of music. The samples are stored in a soundmap and the programme jumps between the samples, playing them in different sequences so that they are heard as different pieces of music.

Obviously, the synchronization of the audio needs to be very precise for this technique to succeed. It works better with music that can be precisely specified, such as that often purpose-written for use with computer games.

Synchronizing sound effects

Images that need a sound to go with them, such as the bang of a gun or the thud of a ball against a bat, can best be served by retrieving the sound from the disc and storing it as a soundmap in the player's memory, to be played immediately a trigger in the video sequence is fired.

Speech

Except for audio-visual segments in which both sound and vision are encoded and decoded together, it can be difficult to link up sound and vision in CD-I programmes. Lip-synching speech to the movement of actors' faces can be done but is a

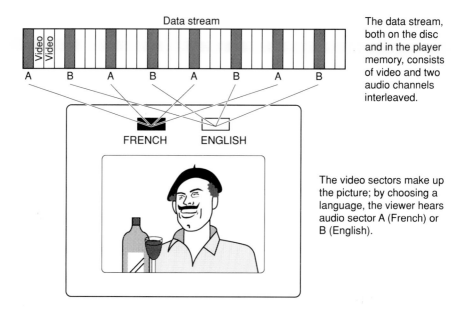

The data stream, both on the disc and in the player memory, consists of video and two audio channels interleaved.

The video sectors make up the picture; by choosing a language, the viewer hears audio sector A (French) or B (English).

Figure 6.1 Interleaving is the alternation in the data stream of information relating to different video pictures and audio tracks, which are simultaneously available to the viewer.

tricky process. Where possible it is much simpler to use voice-over. This has the advantage that several different soundtracks can be recorded for selection by the viewer. The disc could be internationalized by using different languages, or different types of information could be recorded on separate channels for older and younger viewers.

Interleaving

CD-I stores each type of information – audio, video, text and so on – in separate sectors on the disc (see Figure 6.1). The way the different sectors are arranged affects the way information is retrieved and played back (see page 118 for more information

	Relative Sector Number																
Level	0	1	2	3	4	5	6	7	8	9	10	11	12	13	14	15	16
A stereo	*		*		*		*		*		*		*		*		*
A mono	*				*				*				*				*
B stereo	*				*				*				*				*
B mono	*								*								*
C stereo	*								*								*
C mono	*																*

Figure 6.2 Audio sector interleaving. Sound tracks of different quality have different bandwidth requirements: thus level A stereo requires twice as many sectors per second as level A mono and four times as many as level B mono. Within each channel recorded information must be placed on the disc at fixed intervals. This table shows how the information for the six different quality levels is distributed in a sequence of sectors, arbitarily numbered from 0 to 15. The pattern repeats from sector 16.

on disc layout). This process is called interleaving and offers important creative possibilities; it is what enables CD-I to be a multimedia format.

Interleaving is an important facility for the designer to consider because, like the use of video planes, it gives added scope for creativity. Interleaving is particularly useful for:

- Combining different standards of audio on a single CD-I disc. For instance, a designer might want to put music and high-quality special effects on one track, with commentary on another (lower-quality) track.

- Running several parallel audio presentations against a single video presentation – as is the case when there are multiple language commentaries

- Achieving seamless cuts from one soundtrack to another.

The main restriction on interleaving is that sound sectors must be placed in predefined positions to ensure a smooth soundtrack. The placing is dependent on the sound format being used (see Figure 6.2). However, the organization and arrangement of sectors will in most cases be handled by authoring programs and software tools, and designers should concentrate on making sure that their plans do not cross the limits of available bandwidth.

Conclusion

CD-I provides a selection of sound formats to ensure that most design needs can be met. Any kind of sound from the classical symphony to the plainest of computer beeps can be made without wasting system time and performance.

Balancing creative demands with system capabilities is the key to good CD-I design. The right type of sound encoding must be chosen and the sound track must be placed correctly on the disc if audio is to make the best contribution to the success of your programme.

7

The CD-I system

Introduction

From oil painting to radio, knowledge of a medium's physical characteristics is essential for good design. This is especially the case with CD-I: the complexity of the medium demands great precision in its handling. The first task of designers is to familiarize themselves with the performance of the basic CD-I player and disc, and gain an understanding of how physical performance affects design.

The main thing to remember about CD-I is that it is a digital medium. In other words, everything – from moving pictures and music to text and menus – ends up as zeros and ones on the disc. The player decodes it and replays it as sound and image. But this imposes limitations; only a fixed amount of information can be processed at once. This is the bandwidth or data rate of the system. A road junction has a set capacity, and if more vehicles try to use it a traffic jam will result; similarly, data can only be transferred through the player at a set rate that cannot be exceeded. Another analogy is the bucket and the funnel. You may have a very large bucket (and the 650-megabyte storage of a CD-I disc is quite large) but it can only empty through the funnel – the playing device – at the rate set by the narrow neck.

You need to know the dimensions and capacity of the player to know what images and sounds you can present, and how fast you can change them round. You also need to know how much data you can get on a disc, and what that data represents in

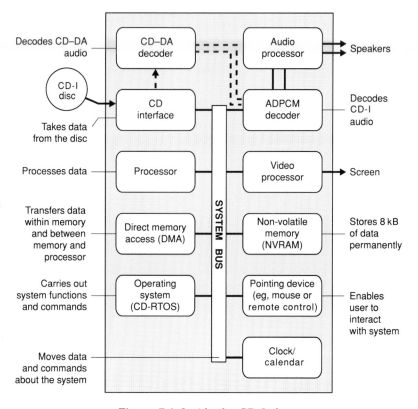

Figure 7.1 Inside the CD-I player.

terms of minutes of music or pictures, or pages of text. Once you have a feeling for these capacities, you can start to see what is and is not possible.

Characteristics of the CD-I player

How the systems inside the CD-I player work is less important than understanding *what* they actually do. The player includes decoders for video and the two types of audio information

(CD-DA and ADPCM), and a controller for accessing the disc and selecting information from it to decode.

Figure 7.1 shows the main structure of the player, how the data comes from the disc and how it ends up as images on the screen or sounds coming from the speakers.

The important characteristics of the player are memory, data rate and seek time. Looked at in computer terms, a CD-I player can deliver data about five times as fast as a floppy disc drive, but it takes about ten times as long to perform a substantial seek operation.

Memory

The player's memory governs the amount of information it can hold 'live' to process and send out to display (Figure 7.2). The working memory of the player is one megabyte; an additional half-

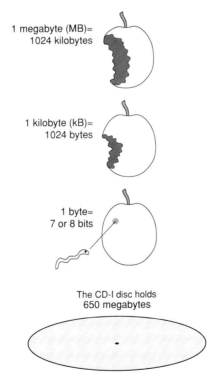

1 megabyte (MB)=
1024 kilobytes

1 kilobyte (kB)=
1024 bytes

1 byte=
7 or 8 bits

The CD-I disc holds
650 megabytes

Figure 7.2 Units for expressing quantities of data.

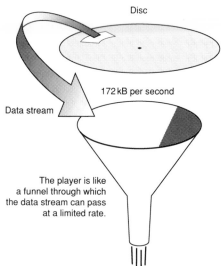

Figure 7.3 The speed at which data can be read from the disc is limited by the bandwidth of the player.

megabyte of memory (512 kilobytes) is dedicated for the use of full-motion video images. This means that at any one time the player can store one megabyte of images, sounds, instructions and so on. The memory is mainly used for storing pictures. So if you use effects that involve rapid changes and high-quality images, the size of memory is very important. A whole megabyte may seem a lot but one frame of a normal television picture takes up about that amount of space. A natural-colour still image (such as a photograph) takes up about a tenth of a megabyte (approximately 100 kilobytes).

Data rate

Data rate, or bandwidth, is the speed at which the player can receive, process and send on information. In CD-I terms it is how much data the player can deliver in a second. The data rate of the CD-I player is 172 kilobytes a second (Figure 7.3).

You can store up to 16 parallel sound tracks on the disc through the technique known as interleaving (see the previous chapter). The data rate is sufficient to allow the player to access all of these simultaneously, so they are all instantly accessible.

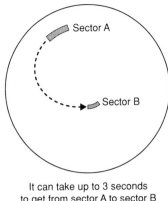

It can take up to 3 seconds
to get from sector A to sector B

Figure 7.4 The seek time of the CD-I system is the time taken to move from one sector to another.

For example, you could use each track for a different language, and allow the CD-I audience to switch from language to language instantaneously. However, in practice you would probably not use more than eight channels in this way.

Seek time

Seek time is the time the player takes to switch from reading one part of the CD-I disc to reading another part. This is tied up with the geography of the disc (that is, where all the sectors are) but the maximum seek time taken to move between two widely spaced sectors is three seconds (Figure 7.4). Sectors are discussed further in the section on CD-I disc layout below (page 118).

Minimizing seek time is an important part of good CD-I design. One proven technique is to design the disc so that sectors containing related data are situated near each other (in coherent design units called modules – see page 119). This reduces seek time considerably and may mean that there are no visible delays in retrieving information.

If a delay is unavoidable, you must let the viewer know that the system is in the process of fetching the information – by displaying a special cursor, a text message or, ideally, a screen designed to look like an integral and functional part of the programme.

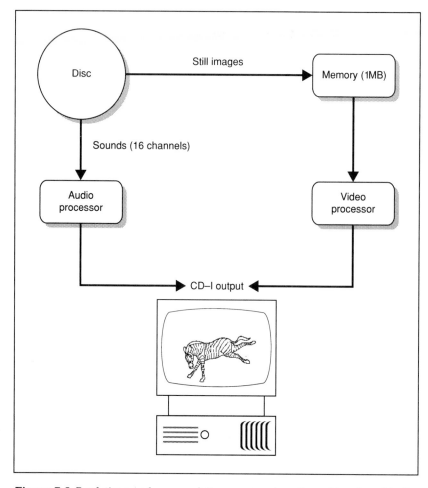

Figure 7.5 Real-time and non-real-time processing. Sound is relayed to the output as it is read from the disc – that is, in real time. Still images are held in memory until required.

CD-RTOS

Although it is intended to be operated as a consumer electronics device, like a VCR, the CD-I player actually contains a powerful computer, the abilities of which can be harnessed by CD-I designers and producers. Detailed technical information on CD-I systems and operating software can be found in the Green Book and other manuals that accompany development systems.

The operating system used by CD-I players is called CD-RTOS. An operating system is software stored in the system and used by it to handle all information from memory, discs and user input, and to process any requests made by the user. CD-RTOS has been specially developed to support the needs of CD-I players and programmes, handle audio-visual material, and monitor the whereabouts of the cursor without using up too much of the system's performance. In technical terms it is known as a real-time operating system, because it responds to user requests at the time they are made (Figure 7.5).

Real-time interactivity is one of the big pluses of the CD-I system. The viewer acts, and the system responds: it gets data from the disc or memory and displays it on the screen in real time – that is to say, as nearly as possible instantaneously. The ability of the system to respond like this depends upon whether the designer has taken into account all the factors described above – that is, upon whether changes between images are paced properly and the organization of data on the disc is well thought out.

Non-volatile memory

The base-case CD-I player has a small permanent memory that keeps useful information even when the machine is switched off. This memory can be used to hold information about the user – for example, in a fitness programme it might hold a small amount of data about users' fitness levels, which saves them having to supply the same information every time they use the programme. It is known as NVRAM (Non-Volatile Random Access Memory), and it holds eight kilobytes of data.

As an optional extra on some extended CD-I systems, there is also a 'personal memory device', which slots into the front of the player and can be used when an application or programme requires larger amounts of information that have to be stored permanently.

The disc

A CD-I disc holds 650 megabytes of data. To anyone used to personal computers that sounds like a lot. After all, a high-density 5.25-inch floppy disc holds 1.2 megabytes, and that seems adequate for many business computing purposes. But you have to remember that converting analogue images, such as the sound of an orchestra or one frame of a movie, is expensive in terms of data storage.

If you convert pictures and sounds straight into digital form, the disc's capacity of 650 megabytes might start to seem limited. After all, one frame of a normal television image takes one megabyte of data, and to make a smoothly flowing moving image, you need about 25 frames per second. Fortunately CD-I's compression techniques massively reduce the amount of space needed to store each image on the disc.

Compression gives digital media a distinct advantage over other methods of recording and storage. There are several ways of compressing both video and audio images so that you can get more on a disc. Each of these methods results in a certain tradeoff of space for definition. For example, you can compress the data for moving colour pictures to a tenth of its size, but it may result in a slight coarsening of outlines and in the gradation of colours. However, while a field of waving corn might look a little less alluring, the compression process would suit very well the bolder, simpler outlines and colours of a cartoon. Compression should not be seen as a compromise. Instead, each compression process should be seen as a way to maximize the virtues of particular types of picture or sound.

Using disc capacity to the full

Firstly decide which media you intend to use. Naturally, you may need to change or amend your decision when you look at practicalities such as disc capacity. You may have to juggle the

Table 7.1 Maximum amounts of material that will fit onto a CD-I disc. (Most programmes will contain a mix of these media and so will not use the maximum amount of each type.)

Full-screen full-motion video (including Level C sound)	72 minutes
CD-DA sound	72 minutes
Speech-quality sound	19 hours
Animated pictures	72 minutes
Graphics	120,000 screens
Natural images	7,000 screens
Text	100 million words

mix of media or change the basic design in order to use the space to the best advantage. These are all processes that form part of the skill of CD-I design.

Tradeoffs

Use the tables in the Appendix to work out how much space your design will take up. You can then decide what compromises, or tradeoffs, will be needed. If the design won't fit onto the CD you have three options:

- scale down the design
- change the format and decide to publish a sequel/companion disc
- leave the design intact and juggle the media mix until it fits the disc

If, on the other hand, the design uses up very little of the disc, you need to decide whether that is because the design is not fully realized. But do not feel that you have to use up all the disc space simply because it is there. The important test is whether you are presenting a satisfying programme to the audience, not whether every last kilobyte of disc space is full. Many excellent programmes use a relatively small amount of disc space.

Table 7.1 shows you how much of any single type of material you can get on a CD-I disc. This is the palette you have to work with. Of course, it's hardly likely that you would use only one medium in this way, but it's useful to know the upper limits for each of them. Also, remember that the audience can play an interactive disc for far longer than any of the individual times shown, because the same resources are used and reused in a variety of contexts.

Laying out the disc

This section discusses a vital part of good CD-I design, laying out the disc itself. In practice the finer details of this process are handled automatically by authoring systems, or managed by the title's programmers. However, the programme designer should be aware of the implications of the way discs are laid out for the way the programme will appear to viewers.

This can have a huge impact, because of the way the player reads the disc. With a traditional linear medium such as video or audio tape, you need never worry where information should be placed. It always goes after the thing before it, and if viewers or listeners want the information in a different order, they must search for it themselves using fast-forward and rewind controls.

However, interactive media are by nature accessed almost randomly, and you will need to ensure that transitions from one area to another can be achieved as seamlessly as possible. Your programme structure will show you the moves that viewers can make through the programme, and you may be able to juggle all the disc information so that the player can find anything the viewer can request without a perceptible delay. To do this you need to understand how data is stored on the disc.

Tracks and sectors

On any CD-I disc, data is stored in tracks of sequentially recorded sectors that contain about two kilobytes of digital information (Figure 7.6). Each sector begins with information

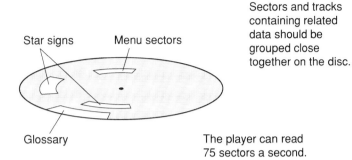

The disc is laid out in tracks made up of sectors. Each sector contains data of one type: video, audio, text or control.

Star signs   Menu sectors

Glossary

Sectors and tracks containing related data should be grouped close together on the disc.

The player can read 75 sectors a second.

Figure 7.6 The arrangement of sectors on a disc.

telling the player what type of data it contains, so that it can be decoded by the appropriate processor. Data types include audio, video, text and software.

Sequences of sectors should be grouped into **modules**. A module is a functional unit of the CD-I program. For example, in a language-learning programme a module might correspond to the section called *In the Café*. It is unlikely that material from *Buying a train ticket* will be called upon when using this module, so the two can safely be stored at different places on the disc, and the possibility of the user experiencing a seek-time delay is avoided.

One of the secrets of good CD-I design is to group tracks and sectors together on the disc in short modules that reflect the way the audience uses the programme. However, implementing this will probably be handled by technical staff, programmers or even automatically by a CD-I authoring tool.

Data transfer channels

Data transfer channels are the paths through which data gets from the disc to the audience. The data stream is the flow of data through these channels. There are 32 channels, up to 16 of which can be used for audio information. As discussed in the previous chapter, most types of audio require more than one channel at once. For example, if you are using basic speech-quality sound, you can have up to 16 channels at the same time; whereas if you are using level A stereo, you can fit in only two channels at the same time.

The CD-I player combines video, audio and other data that has arrived in the data stream through a number of channels, and combines the data they carry to make up the multimedia mix required by the CD-I programme. Any mix of data from these channels can be used. Information from them is interleaved to make up the programme that the audience sees, and from which it makes choices and selections.

For example, a travel guide might include voice-overs in six languages, along with background music and photographs. All this, along with control information, is carried in the same data stream that is coming from the disc. This means that the audience can change from one language to another instantly, without interrupting the pictures or the music. The player does not have to go away to another part of the disc to find the parallel language commentary.

Working out your programme's demands

The previous sections have carried a large amount of technical information. This section gives you a basis upon which to start looking at what materials and effects you are going to use, and how you are going to use them.

You could make a table like Table 7.2 to assess the technical viability of your programme idea. By making a table of this sort you can start to have some idea of whether your CD-I idea is technically feasible or not.

Table 7.2 Time and space requirements for a projected CD-I programme. (The total memory requirement for one disc must come to less than 650 megabytes.)

Visual materials	Type (CLUT, DYUV, etc.)	Time used	Space used
Full-screen, full-motion visuals			
Other types of moving video			
Animation			
Still pictures			
Graphics (menus, etc.)			
Text			

Audio materials	Level (CD-DA, level A, etc.)	Time used	Space used
Speech			
Music			
Sound effects			
TOTALS			

Other technical issues:
- Am I going to use NVRAM? YES/NO
- Am I going to use soundmaps? YES/NO
- Am I going to use two picture planes? YES/NO

Will it work?

Find the place in the programme where the use of resources is most intense, and calculate whether the CD-I player will cope with the data stream.

Let's say that the producer wants a fast-changing sequence of full-screen pictures for 10 seconds or so, with level A stereo sound. Each picture is required to cut to the next one every second. For this example we will ignore any implications for the player's memory. Full-screen DYUV or CLUT images take about 100 kilobytes or 45 sectors each (accurate enough for this example). The data rate for images in this sequence would be 45 sectors per second.

However, the level A stereo sound takes half of the total data bandwidth of 75 sectors per second, leaving 37.5 sectors per second for the images. This is insufficient for the proposed sequence. The solution is to trade off. The following are possible options:

- use less than full-screen pictures
- use run-length coding compression if the pictures are suitable
- reduce the sound quality level by one step, to level A mono or level B stereo
- change the pictures at a lower rate – say, every two seconds.

Any of these alternatives would make the sequence possible within the planned time. Which option you adopt is determined by the tradeoff process. An agreement has to be reached by the project manager or producer and the design team as to which option should be adopted. Such a process is of the essence of CD-I design.

Conclusion

Understanding the way the CD-I player and its discs work is vital for good design. It's also important to be aware of technical considerations from the initial design stages so that, when it comes to production, resources are not wasted by the need for last-minute shuffling of material or by the production of sound or video recordings that will never be used, owing to lack of disc space or bandwidth.

> ***Final design checklist***
>
> *Will everything I want to include fit onto the disc?*
>
> > *(If not, will I trim down the material or go on to a second disc?)*
>
> *Will all my effects work on the player? (That is, will they 'go through the neck of the funnel'?)*
>
> > *(If not, how will I redesign them - what sacrifices or tradeoffs am I prepared to make in order to make the idea work?)*
>
> *Will the production use compatibility mode, or will it be specifically PAL or NTSC?*
>
> *What qualities of sound and image will I be using?*
>
> *How will I use the visual planes?*
>
> *What effects will I use for moving between screens?*
>
> *Will I be using soundmaps?*
>
> *How am I going to break down my programme idea into modules that can be arranged in geographical proximity on the disc?*
>
> *How can I use the channels available, if I need to, to present parallel soundtracks?*

Designing the physical layout of the disc is almost as important as designing its creative content, in terms of ensuring that the viewer is presented with a compelling and believable programme to watch and interact with.

8

The design process

Introduction

This chapter discusses the stages of the CD-I design process. Every CD-I title has its individual content and media mix and therefore its own individual design process. Nevertheless, it is possible to establish stages that are common to most design projects. The design process model shown here can be adapted by designers and project managers or producers to fit the requirements of their particular CD-I title.

Stages of design

The design process can be divided into three main stages (Figure 8.1). At each stage, tasks can broadly be divided up into **design** and **management**. Design tasks are those that are concerned mainly with the eventual production of the detailed design specification. Management tasks are concerned with managing the creative process and linking the process to the outside world – whether it be an in-house production department, or an external client or production house.

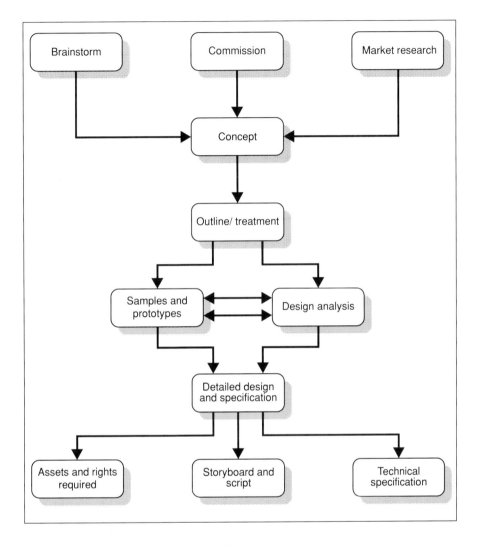

Figure 8.1 Typical CD-I design process.

The design process is itself an interactive one, with information passing constantly between design and production specialists: it is the job of the producer or project manager to facilitate this process. In most cases the design will be the responsibility of a team, which includes technical and line management staff as well as programme designers and content experts.

One aspect of CD-I design that almost all designers agree on is that a multidisciplinary approach is vital. It's never worth excluding technical staff from even the earliest conceptual and brainstorming meetings, because their informed input is vital. Another point is that the design stage must have enough resources devoted to it. One designer reckons that spending a quarter of the budget on programme design is a good rule of thumb, although titles that require ambitious work in the production stages may not be able to meet this.

Stage 1 – Concept and treatment

The very first element in CD-I design is discovering the needs of the audience for a particular title. That knowledge, together with the creative spark that may have preceded it, are developed into the programme concept, which outlines the purpose of the CD-I and the way it will achieve that purpose.

The treatment is like an artist's sketch that includes notes on how a painting can be made from it. To the baldly stated programme concept it adds more details on the way the creative aspects of the programme will be worked out.

It will often be supplemented by a prototype, a more detailed mock-up of a part of the planned programme. How you present your concept and treatment, whether you include a detailed prototype and so on, will probably depend on whether you are producing a CD-I in-house for an existing producer, or trying to attract funding from a company new to interactive media.

A CD-I designer explained how one programme developed from a very simple concept: 'We had the idea of caricatures, distortions of people's faces. We used design tools to distort photographs, and then realised that we could use CD-I's video planes to allow viewers to draw over photographs.' The design was developed in two ways from this initial concept: the tools for drawing the caricatures and the supporting software needed to make it work had to be defined, and the programme content, the images and photographs to be used, had to be listed and organized in a format suitable for CD-I.

Stage 2 – Design analysis

In the second stage the parties concerned determine just what the proposed design will require in terms of materials, software and legal and administrative arrangements. The design analysis is like the recipe listing the ingredients for the production. It may draw on experience gained while working on the prototype; on the other hand, some CD-I designers may prefer to produce a prototype at this stage.

Stage 3 – Detailed design

At this stage the design is specified in great detail and should also be validated – tested to make sure that it will work as planned. This is a process by which technical staff such as programmers look at the design ideas and make sure that they are achievable. The design may need to be changed slightly to make production easier. Alternatively, one production path may be chosen instead of another because it enables the design to be more fully implemented.

The three stages are discussed in detail in the following chapters. In addition, each chapter includes lists of the documents and other items that should result from it.

Depending upon the degree of formality with which the project is organized, each stage can be 'signed off' before proceeding to the next stage, or the next stage can commence when everybody feels about ready. Some people see a very formal framework of organization as an impediment to creativity, while others see it as an essential component of a well-run project.

Production stage

Once the design has been finalized, the detailed design document is usually signed off and contracts signed for the production phase. Timing and project management now become important, as the various pieces of production work need to be coordinated and dovetailed.

The CD-I design team

'It might be possible one day to design a CD-I programme without teamwork; it can be done in some media. But right now I don't think it can be done that way. There's no one who thoroughly understands all the different disciplines,' was the opinion of one experienced interactive designer.

There are as yet no rules about the best way to organize a CD-I design and development team, but good practice suggests that the design team contains a representative from each of the disciplines involved, who will then go on to manage production teams working on their own areas of the programme's development. The team for any given project may need to include:

- graphic designers (including animators)
- programmers and other technical staff
- writers and researchers with expertise in the content of the planned programme
- video and sound production staff for any new material that needs to be recorded
- project management staff to keep the project on time and in budget
- creative management staff to make sure that the design is fully developed and carried out

The number and type of staff needed will obviously vary from project to project, but the kinds of skills needed will stay the same. Most CD-I designers agree that, while it might be tempting to exclude programmers from early stages of design development, it's a good idea to have someone around who thoroughly understands the technical aspects of CD-I and can quickly spot any potential pitfalls before effort has been wasted on designing something that can never work. A CD-I manager related one case:

'When we first got involved in CD-I our designers spent a lot of time brainstorming, and came up with what they thought was a brilliant concept for an interactive programme that would teach viewers how to edit videotape and put together their own movies from clips on the CD-I disc. It was a great idea, but when the technical people looked at it they knew it couldn't work, because of the way data is stored in fixed places on the disc. Viewers would never be able to get instant cuts, and with their chosen film clips all over the disc their "movies" would be full of long gaps for disc access. Now we always have a technical representative in meetings for creative staff.'

The design manager

With so many disciplines contributing to the success of CD-I programme-making, one question remains: who should be in charge of a project? CD-I breaks down some of the traditional barriers between different kinds of design and production skills. But clearly someone needs to have an overall view of what is going on throughout the whole project. Who this person is, what they call themselves and what their brief is, very much depends on the kind of CD-I being produced.

On one hand a programme-maker from a media background might be called the producer and use techniques learned from successful programme-making. This kind of person might assume an important influence in the creative decisions.

On the other hand, a team leader from a software background might be called the project manager. He or she might apply software development disciplines, such as project milestones, and tend to leave artistic decisions to others. One CD-I developer says:

'In our studio the project managers tend to be the people with the strongest background in interactive design. It's interactivity that makes CD-I different and it's important to keep that at the forefront in every project.'

Whoever does it and however they are described, managing the creation of CD-I design is an important and demanding task. An important part of the job is facilitating communication between the various work areas and activities in the team, and synchronizing and coordinating their efforts.

The design team

Designers working with CD-I are likely to come from a range of backgrounds and to bring a variety of experience to the project they are working on. You may prefer to use designers who are already confident with the development platform on which the CD-I will be produced, as there will be less time spent on learning how to use the computers. On the other hand, you may want to use a designer whose traditional skills have not yet been tested on the new electronic media. Try to ensure that there is always someone on the team who is an expert on the design tools that are being used.

Ideally all the members of your team will be familiar with at least some aspects of the technology they will be using, and with each other's disciplines, so that, for example, the programmers and designers can understand each other. It is important to discuss issues and exchange information because in a multidisciplinary team operating in a new area you cannot assume that everyone has the same idea of the process, or of their part in it.

Feedback

In the whole of the CD-I design process, feedback is an essential way of discovering what is really (a) desirable and (b) possible. Try to focus the feedback from the review of the treatment. For instance, people may be confused by a paper-based summary of proposed interactivity. The paper version may look very complex to them. It may be a good idea to develop a limited example on a computer in order to demonstrate the flow of the programme and the way in which its logic works; see 'Making a sample module or prototype', page 137, for ideas on how to do this.

However, although this has the advantage of showing some aspects of the programme clearly, you should also point out that a prototype or model may not represent the look or feel of the programme very closely. Understanding this limitation, the clients will be able to read the rest of the treatment more easily. Their reactions will be more informed and therefore more useful.

If you are circulating your treatment to many people, it may be an idea to prepare a form or a questionnaire for their comments. This helps people who are not sure what to comment on and who therefore make unhelpfully general remarks, or say nothing at all. The form may be distributed as part of the treatment so that respondents have some guidelines to follow. This approach is probably more useful for CD-I development within one company; if you are submitting a proposal to another company you will probably want more and different feedback than a form can provide.

When you come to validation and testing of the finished programme this approach will prove even more useful in tracking down problems with the way the programme works.

Conclusion

Design methodology for CD-I is evolving quickly in response to the experience of designers and producers already working in it. While each studio will develop its own way of working to suit the projects and budgets it has to deal with, some points of good practice have already emerged.

The design process	Design	Management
Stage 1: Concept and treatment	Concept Treatment	
Stage 2: Design analysis	Design analysis document Overall flowchart Interactivity script Script synopsis/outline Storyboard Technical estimates Constraint analysis (tradeoffs) Sample module Flowchart Interactivity Script/storyboard	Production path schedule plan Budget breakdown Team list Legal and other matters Agreement to proceed
Stage 3: Detailed design	Detailed design specification Detailed flowcharts Full interactivity script Detailed script/storyboard Inventories (eg: lists of audio, video and and other materials) Final technical estimates	Production path path schedule Detailed budget and cashflow forecast Final team list Full documentation of legal and other matters Contract
Production	Producing original materials Converting existing materials Creating software engines Disc-building Emulation Mastering Validation Replication Packaging and distribution	

Working as a team is particularly important for developers new to the medium. CD-I design requires such a broad base of knowledge that no one person can be expected to acquire it all instantly. Ensuring that the team includes someone with a firm grasp of the technical side of CD-I can avoid many pitfalls when it comes to production.

Expanding the design from initial concept to a detailed specification can be done in stages as other aspects of the production become clearer. There's no point moving on to the detailed design stage while the treatment is being considered by the finance department for budget allocation; better to wait until the resources available are more certain.

What is always good practice is to involve all the team and people beyond it, such as test audiences, in a carefully regulated feedback process. This is particularly important when a title is designed for a specific audience, such as children, whose needs and reactions to a title may not be obvious to programme designers.

9

Design analysis and prototyping

Introduction

Moving from the initial programme concept and treatment to a finished script ready for production is the core of CD-I design work. But it can also be difficult and frustrating when the medium is new to the designers and the level of detail and materials needed are unclear. Maybe financiers have to be persuaded that a project is viable and worth funding fully. Maybe you as the programme designer are not sure about the amount of information you plan to include on each screen.

There are many reasons why you will want to create a prototype of at least one sample module of your planned programme. This chapter looks at ways in which this can be done, and at how to break your overall design into manageable sections. It also covers the choice of design tools and staff to carry out the design.

Planning the design

Turning a programme outline, concept or treatment into a detailed specification that is ready for production can be done in many ways. What all approaches will have in common is that the result will be a document of some kind. However, just like the programme treatment, this document can take many forms and need not be a printed text. One experienced CD-I designer says:

> 'Traditionally, everyone says it's critical to have these documents, but sometimes people can't understand what you're talking about when you use words to express it, especially in relation to interactive design. Designers need to communicate the planned look and feel of the programme, and the traditional hierarchy of documents may not do this. The alternatives don't have to be very sophisticated – a series of drawings, or a simple HyperCard stack – but they're much more useful.'

Making modules

Try to see your programme in modular terms, and identify the common elements of each module. This has a decisive effect in reducing production time for the design, scripting, visuals and software. It breaks down many tasks into manageable chunks and shows where design effort need not be duplicated. For example, a visual encyclopaedia of marine life could be divided into modules based on each entry in the main menu, which divides sea life into different biological categories. The structure of each module is the same as far as the programmers are concerned, but the picture researchers can be handed a section each, while the graphic designers add the appropriate fishy touches to each menu.

Looking at your design in this way should also help you to work out the budget required, and see whether your plans need to be scaled down or expanded. If you have not already got your design team in place it may help to show where you need more help – whether another animator is needed, or a scriptwriter, or a picture researcher.

Splitting your programme into modules will have technical benefits when it comes to production of the disc, as well as simplifying programme development. A sound underlying software design with a logical structure greatly speeds up CD-I development. In particular, the use of software engines within one title and across a series of titles can save a great deal of money. This is one reason why it is important that the person in charge of the project should have a good understanding of how CD-I works.

Choosing development tools

As you begin to work out details of your design and the path along which you will travel to complete it, you will need to consider the design tools that you and the designers will be working with. CD-I supports a wide range of tools for development, design and prototyping and if you are new to CD-I you should attempt to evaluate as many as possible. Do you want to use dedicated CD-I development and design software, or would you prefer to stick with a multimedia authoring tool that you already know, which can output its files in CD-I format? The choice of authoring and design software will have a major impact on your working environment for some time, so it's worth investing some time in making the right choice.

Even if your development system is already in place, it's worth checking to see whether since your last project there have been any updates to the software you plan to use, and keeping a look-out for any new development tools that might be easier to use or provide facilities not present in your existing system.

While your initial preproduction design work may be done on a completely different system to that which will be used in production itself, you should be aware of any development systems that will be used in production and ensure that the design is tailored to their capabilities. For example, one type of dissolve between images may work much better than another on the system you will be using, or the software may not have a predefined paintbrush tool that you were expecting.

Keeping track of design information

Managing the flow of data in a big CD-I project may sometimes be difficult. Large amounts will be generated during the project, much of it computer-based, so design processes and data management processes must be applied in tandem. The management implications of design decisions and the design consequences of management decisions should always be considered.

Logging and backing up data

If your development platform doesn't do so automatically, it may be appropriate to set up a small database and log each piece of video, each audio section and so on. They can each be given a unique identifying name that will help the team keep track of them throughout the project. While you're logging your data, make a back-up copy of it on a spare disc and keep it safe against the inevitable computer disasters.

Naming files

It is a good idea to establish naming conventions for files held on computers right at the start of the project. Even files that are usually used only by one person should follow a convention understood by everyone on the project, so that no-one has to waste time searching for lost files.

Making a sample module or prototype

Visualizing the way an interactive programme will appear in its final form can be very difficult, especially for those with limited exposure to multimedia. A paper-based description can't really do any programme idea justice, and you may find that developing a prototype or a sample programme module is the best way both to understand CD-I's capabilities and to explain them to other people not immediately involved in the design process.

Prototypes and sample modules

You may even find it useful to create *both* an overall prototype, which is a relatively rough mock-up of the whole or part of the planned programme, *and* a sample module, which is a small part of the programme produced in the same way and to the same standard as the real thing will be.

A prototype of the whole programme can help you to decide whether the structure of the programme works properly, whether it goes at the right pace and whether the pace is consistent. It can alert you to major content omissions and can test design details such as overall programme metaphors to see if they really work 'live'.

A sample module can give other valuable information, helping you to assess whether the chosen graphical style is right; it can give you further insight into the pace of the programme; and it can give outsiders an even better picture of the way your programme could work if developed in full. It also provides an opportunity to test the way your team works and to develop an efficient way of working.

What goes into a sample module

The project manager or producer can bring the different design processes together to create a small sample module, including (as appropriate):

- text
- graphics
- interaction
- video, audio or simulations
- logic

The sounds and images should be as close as possible to the quality of sound and image that you plan to use once full-scale development goes ahead. You should develop a large enough sample to cover all the options offered by a menu in the programme, so that you can test movement between items and levels of information. The graphics used should also be close to those you plan to use in the actual programme, although you may be testing alternative approaches.

What goes into a prototype

An integral part of the detailed design stage may be the creation of prototypes. The project manager or producer has a coordinating role in the creation of prototypes, since these involve input from several members of the design team.

Prototyping typically involves testing some of the design options. For instance, different screen fonts or forms of menu text might be tried, to see which is most suitable. Or you may be testing the whole programme structure to see if you have got it right or if the material needs to be rearranged for ease of use.

It is helpful if prototyping also involves a user, preferably one with the same level of expertise as the intended audience. The design team can assess whether a prototype 'works' in design terms, but it is far better to allow a representative of the audience to assess whether it 'works' in terms of usability.

When to create a prototype or programme sample

It's likely that a prototype of some kind will be needed in most CD-I projects, but they will be needed at different stages of development according to the nature of the project. The most useful time to develop a prototype is after finalizing the programme concept and treatment, when you know what the programme is about and how you intend to approach the

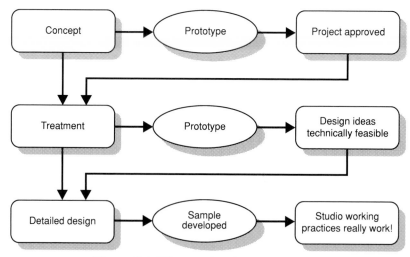

Figure 9.1 When to develop a prototype.

material, and before the detailed design stage, in which every last screen, word and image is laid out (Figure 9.1).

One CD-I designer created two prototypes for a large project to design a multi-player game that would be like an interactive version of a traditional board game:

> 'Two aspects of the programme needed to be tested. We didn't know what it would look like, so we created a quick animation of several screens using the MediaMogul authoring tool. We didn't know what it would be like to play, or how to get the right mixture of skill, judgement and look into the game, so we developed a logical model of it on a PC. Although it's only text, it's quite sophisticated, and after a lot of work it's now fun to play. We're hoping that we will be able to transport a lot of the model into our CD-I authoring tools to use as the basis for the programme itself.'

However, a prototype may be needed to secure funding for a programme that is only at the conceptual stage; in this case it will be relatively vague and will need to be supplemented by programme samples showing ways the programme could look. The relative ease of design development afforded by good authoring systems should encourage you to develop prototypes and samples to experiment with all the possibilities that CD-I and multimedia offer.

How is a prototype developed?

Prototypes are usually developed on a computer that can use its hard disk to simulate the action of a CD player – for instance, Philips' Starter Kit. Another option is the Apple Macintosh hypermedia programme HyperCard, or similar programs designed for other computer systems. At a more advanced stage of the project, prototype discs may be cut on a WORM (Write-Once, Read-Many) drive. This is a drive that allows you to make CD-I discs relatively cheaply without going through the mastering process.

Prototyping is a process of creative exploration. However, bear in mind that a prototype is not a miniaturized version of the final thing. It demonstrates the way the title works, but it is not necessarily an accurate representation. If you present a proto-type to clients or customers, you should be careful to explain that this is the case. You should not present the prototype if you feel there is room for misunderstanding on this point.

Also, remember that if you spend large amounts of time and effort on one prototype there will be no budget to try out others, or to make fundamental changes. The prototype is a tool to help the design process. It should not take on a life of its own, dictating the way the title has to be designed. Testing, prototyping and developing samples – all of these should feed into the design analysis as part of a larger iterative process.

This does not preclude designing the prototype to be as reusable as possible. If you are prototyping a section involving a menu choice, make it as near the real thing as possible (within the constraints discussed above).

Conclusion

Producing a prototype or sample is an easy way to experiment with your CD-I programme ideas, whether it is to generate programme concepts for future use, or to finalize your thinking on aspects of a programme in development.

You should aim to make simple prototypes of as much of your programme as possible, and a detailed sample version of a small section, before finalizing your design. It will prove invaluable in the later stages of the design process, design analysis and detailed design.

By the end of stage 2 (see box, page 132), sample visuals, script synopsis and script samples should have been prepared as well as the sample module. In addition, the software engineer will have developed an implementation specification. After approval, this feeds into stage 3 of the design process. The process continues in a loop until the design issues have been resolved and the prototype is acceptable.

10

Detailed design

Introduction

It takes more than a brilliant concept to make a good interactive television programme. The inspiration needs to be backed up by a mass of detail, so that every image and sound can be sourced and encoded into the correct CD-I format, every menu written and designed without the need for changes, and the programming begun with all team members sure of their own responsibilities.

In stage 2 of the design process, design analysis, the designers take the treatment a stage beyond the initial treatment. The writer or content designer may write a synopsis of the script or storyboard. The artist develops sample visuals. The software engineer assesses the technical viability of the design, including estimates of how much disc space it will take up. The engineer's role is to recommend the appropriate software tools and to begin to outline the logical structure needed on the disc. This chapter looks at the design input needed at this stage.

Once the outputs of the design analysis stage have been agreed and signed off by everybody involved, the design has to be closely specified before production can commence.

If design analysis is like making a shopping list before cooking a meal, detailed design is like writing the recipe, and production is like following it. The recipe or specification should be as complete as possible in every detail so that the people involved in production can pick it up and start cooking.

At this stage, the project manager or producer must ensure the overall cohesion of the design. At the same time he or she must monitor the project, to ensure that both the budget and the production schedule are on target.

Developing the design

Your design will by now have developed into a well-structured outline to which detail is beginning to be attached. Some areas will be highly detailed, because they have formed part of a prototype or sample module that has been developed further than the rest of the planned programme.

The design process will be generating many documents by this stage. Some of them will be conventional paper documents, others will be computer files made with the development or prototyping system. If design information is being split between several documents, you will need to ensure that all are kept up to date with each other. One good way to combine all the information is to keep it as a hypertext file, such as a HyperCard stack, which can hold all the details and accurately represent the planned interactive structure of the programme. Documents at this stage may include:

Overall flowchart

This outlines the structure of the programme. If the programme is well designed, this should be enough to indicate its shape fairly closely.

Interactivity script

You will need to record how your programme uses interactivity, showing where each choice takes the audience to, and how they get from one part of the programme to another.

Script synopsis / outline storyboard

You will need to show in broad terms the plot of the CD-I. The storyboard should indicate all the elements, modules and parts that are planned for the title, and include a brief description of each of them where necessary.

Technical estimates

By this stage you should also be thinking about the feasibility of the title in technical terms – broadly: how much data and processing power do specific parts of the programme require? Will all the data required fit on the disc? Can the processing power of the CD-I player cope with all the effects required?

Constraint analysis

This is a list of the disc's capacities, against which you can map the demands your design is making. The list might include:

- memory
- resource analysis
- disc bandwidth and capacity
- seek performance test
- microprocessor and real-time performance
- efficiency utilization
- designer-specified parameters
- PAL/NTSC constraints and compatibility issues

Most of these issues have been examined in detail elsewhere in this handbook. In particular, you may find the table in the Appendix useful in working out a constraint analysis for your CD-I design.

Design analysis

Analysing the design should produce the results shown below, which will help the project move on to the final stage of detailed design. In some CD-I projects, a complete and detailed design will result from this stage, ready to go into production. In others there may be some questions still to be answered before production can begin.

You should use your established review and feedback processes at this point to make clear what design work remains to be done and what loose ends need to be tidied up before production can begin.

Production paths

Production paths give details of how the various materials used in making the CD-I title are themselves going to be made. For example, if material in analogue video is acquired, how and where is it going to be converted? Where are musical interludes coming from, and will they be in digital format, or will they need converting? The idea of this document is to show how all the materials, and the disc itself, arrive in the right form in the right place at the right time.

The production schedule

To make this, the project manager or producer has to get estimates from all team members of how long their work will take to complete. From these estimates, a schedule can be compiled. The schedule should show:

- how long each process will take
- the order in which the processes should happen
- which processes are interdependent

- which can overlap
- project milestones
- expected release date

Some processes, such as software design, need to start earlier and progress faster than others.

The production plan

Preparation of the production plan is linked to the production schedule. This shows the production dates and routes for all the material that is going to be produced. The project manager or producer has to test all these routes to make sure that the design can be produced on time and to budget.

It is particularly important that the project manager or producer takes responsibility for ensuring that material is produced in a suitable format. For instance, if graphics are to be produced they must be in a file format compatible with CD-I.

The budget breakdown

Preparing the budget is sometimes difficult at this stage because of the lack of detailed information. The script or storyboard is a key document for estimating costs and for keeping costs under control once production begins.

By linking budget estimates to the production schedule it is possible to produce a cash-flow forecast, showing when money will be needed during the course of the project. Remember to include a contingency fund in the budget. This will cover delays, unexpected problems, unforeseen rises in costs and so on.

Legal and other clearances

As with any other commercial project there will be legal issues to be dealt with. Staff contracts are clearly important. Other areas include:

Agreement to proceed

The way in which the agreement to proceed to detailed design and production is expressed will depend on whether the CD-I is an internal development or one that involves a production house. Even internal projects may involve formal arrangements between cost centres, departments or subsidiaries. Remember that CD-I is a new area and be sure when agreeing contracts that all parties are quite clear about the terms being used. For instance, do they all mean the same thing when they refer to 'design', 'development' or 'production'? It may be better to define these terms somewhere in the contract, so that everyone is clear from the outset what is covered.

This formal agreement to proceed may be made at various stages. Occasionally, it will occur before stage 1 (concept and treatment) begins. Sometimes it will not happen until a very detailed design has been approved. Most often the model described in this manual, consisting of three stages, will be followed. Sign-off of each stage initiates the next stage.

Copyright clearances

The project manager or producer is responsible for ensuring that any copyright or other legal matter connected with the disc is researched and resolved. Production studios frequently have information on matters such as rights to music and can help the CD-I developers they are working with.

Rights may seem like a minor issue but, when they are involved, may take up a large amount of budget and time. It is worth investigating them thoroughly at an early stage, if only to make sure that problems that could possibly be generated later on are avoided.

The team list

By this stage you should know your staffing requirements and be able to list the people who will be involved in the project. Bringing the team together is the project manager or producer's job. At stage 2 (design analysis) the team may still be very small

– perhaps only three or four people. But at stage 3 (detailed design) more people will probably become involved. The project manager or producer must identify

- roles
- responsibilities
- relationships
- personnel

An experienced CD-I designer describes the workings of one company in these terms:

> 'The initial idea may come from one person, but the design team will usually have about five members, and that will expand to around 25 people as the title goes into production. Once the bulk of the work is complete, maybe three or four people will work on validating the title and testing it, depending on the complexity of the programme. '

CD-I gains greatly from the fact that it is a multidisciplinary process. The project manager or producer should try to find people who are themselves multidisciplinary: a writer who is also an interactive designer, or a graphic artist who also has experience in designing interfaces. Bear in mind that the team members must be good communicators. The design process is itself an interactive one, with information passing constantly between design and production specialists.

Outputs from stage 2 of the design process

Management documents
Production path, plan and schedule
Budget breakdown with cost alternatives
List of team members
List of legal and other concerns – for example, rights
Contract or agreement to proceed

Design documents
Outline overall design
   Flowchart
   Script synopsis (audio and visual)
   Interactivity design
Technical estimates
Prototype or sample module

The final stages of design

Aim to make stage 3 a self-documenting process. At the end of this stage of the design process, you should have the documents listed in the box.

Outputs from stage 3 of the design process

Design documents
Detailed visual design, shooting script
Detailed graphic design
Fully written content/text script
Interactivity script
Structural analysis and flowcharts
Constraint analysis (tradeoffs)
Software structure
Implementation specification

Management documents
Production path, plan and schedule
Production list
Detailed budgets and cash-flow forecast
Information on rights and legal matters
Contract or agreement to continue

The detailed visual design, shooting script, detailed graphic design and fully written content/text script are all documents that are familiar from other media such as video and film.

The interactivity script

This is not necessarily a separate document, but may be a part of the fully written script. For example, if the script is divided into modules, at the beginning of each module there could be a brief description of the nature of the interactivity in that module.

Structural analysis and flowcharts

These are familiar to those from a software background. If the interactivity has been kept fairly simple (as it should be in a good design) the structure of the programme should be adequately described by a small group of flowcharts.

The constraint analysis

By this stage this might consist of the bare statement of production alternatives. For example, in a certain section it might be a question of stating for the benefit of the production team that there is a choice between a certain sound quality and a certain picture quality.

The software structure

This will show the program design and would, for example, indicate the use of software engines where relevant.

The implementation specification

This describes exactly how the detailed design is going to be carried out.

The production path, production plan, production schedule.

These are tightened-up versions of the documents issued at the previous design stage.

The production list

This document details all the materials or assets required for the production. These are lists of the materials that will go to make up the CD-I. Typically these might include:

- image inventory (stills, video, text and graphics screens)
- audio inventory (sound clips with sourcing information)
- software inventory (list of routines, with engines, software library usage and so on)

The project manager or producer usually takes responsibility for agreeing these lists with the members of the design team. The lists can be cross-referenced against the production plan and the production schedule. They are important both for making sure the various materials can be obtained or created and for costing the production.

Detailed budgets and cash-flow forecast

These documents are now more specific than they were at the previous stage.

Rights and legal matters

The information available on these should now be more definite than at the previous stage, so that production decisions can be taken on the basis of it. If, for example the copyright for a piece of music is forbiddingly expensive, it might be easier to use another piece of music or to make something up in the studio.

The contract or agreement to continue

This is the document in which the client or production house legally commits itself to carry on to the production stage.

Production

Once the detailed design has been finalized and contracts or agreements signed, the production team carries out the design. The materials are developed and the software specification is implemented. Although this is a separate stage in the process, it is also true that production and design in CD-I are closely linked. Many production issues will also involve design issues and vice versa. In the process of realizing the design, the production team will often need to go back to the designers, or to make design decisions themselves. In many cases the producers will be the people who created the original design and are now fleshing out their original storyboard, or supervising a larger team of staff who are doing so.

Some of these decisions will be, for instance, tradeoffs. But given the newness of CD-I technology, it will also happen that effects will have to be achieved by routes other than those originally planned, or even be replaced by other similar effects because of technical considerations.

Obviously, the project manager or producer has to manage this process, ensuring that project targets are being met. Additionally, the project manager or producer will be exercising creative control. This ensures that the creative vision survives during the detailed implementation of the design, now taking place.

Conclusion

By the time the programme design is ready to be put into production, many of the creative decisions will have been taken and the shape of the programme finalized. Technical staff and designers, who may even be the same people, will be able to work together to implement a design that makes the fullest use of the technology by which it is produced and on which it will be played.

Moving from an outline idea sketched on the back of an envelope to a CD-I disc crammed with 650 megabytes of entertainment and information is a complicated process and each team of designers and producers will no doubt implement their own unique methodology. But a process such as that outlined in the last three chapters of this book, which emphasizes a controlled and well-documented progress to the finished object, is recommended for designers new to the complexity of multimedia development.

While the use of computers provides many short cuts that save time from the production point of view, no CD-I programme will be successful if it has not had detailed consideration and careful preparation from its creators and designers. No amount of technical expertise can make up for a programme idea that is devoid of interest or content, or has been produced without any thought for the audience and its needs.

Designers have a vital role to play in the creation and production of CD-I programmes, and in its success as a medium for information and entertainment. No technology can ever usurp the need for creative human input to programmes, and designers can feel confident that their contribution to any CD-I programme is an indispensable one.

APPENDIX

Designer's ready reckoner

This appendix is designed to help you calculate the technical viability of your CD-I design. Two questions you have to bear in mind when designing a CD-I title are:

- does the material fit on the disc?
- will all the sequences work on the player, or do they require too much bandwidth?

In both cases you have to do some careful calculations, as follows.

How to calculate disc space usage

Total disc space is 650 megabytes. To find out if your design exceeds this:
(1) For each sequence, work out the space requirement in sectors. This involves adding together all the material and controlling software.
(2) Multiply the number of sectors by 2.352 to obtain the number of kilobytes.
(3) Repeat for each sequence.
(4) Total the kilobyte figures for all sequences, and then divide by 1,024 to produce the megabyte total.
If the total is less than 650 megabytes, you are OK. Otherwise start thinking about how you can save space.

Will it work on the player?

The maximum rate of data transfer from the disc is 75 sectors per second.

(1) For each sequence, work out how much space in sectors the material it uses takes up.

(2) Divide the total number of sectors by the length of the sequence in seconds.

If the answer is less than 75 sectors per second, you are OK. Otherwise you have to start using fewer or less space-consuming items, or use the player's memory to hold things until they are needed.

Useful figures

Image sizes

- CD-I stores information in sectors that contain just over two kilobytes of user information.

- CLUT 8 or CLUT 7 image figures are the same as for DYUV: between 85 and 100 kilobytes in most cases.

- RLE images vary in size according to the complexity of the image to which RLE is applied, but are typically around 10 kilobytes each.

- If you are not using the motion-video picture plane, you can achieve motion of 15 frames a second by updating one ninth of the area of a DYUV image, or the whole of an RLE CLUT 7 image.

Sound format sizes

With sound it is important to remember the interleaving requirements for real-time playback as well as the bandwidth usage: sound requires not only bandwidth but specified sections of it (see Figure A.1).

	Relative Sector Number																
Level	0	1	2	3	4	5	6	7	8	9	10	11	12	13	14	15	16
A stereo	*		*		*		*		*		*		*		*		*
A mono	*				*				*				*				*
B stereo	*				*				*				*				*
B mono	*								*								*
C stereo	*								*								*
C mono	*																*

Figure A.1 Audio sector interleaving. This table shows how recorded information for the six different audio quality levels is distributed in a sequence of 16 sectors, 0–15. The pattern repeats from secor 16.

System memory

- The player has a RAM memory of one megabyte, which is divided into two banks of 512 kilobytes each. This holds pictures, soundmaps and software. The CD-RTOS operating system uses 50 kilobytes. Material can be accessed more quickly from memory than from the disc, but it is lost when the player is switched off.

- One DYUV image uses about 20% of the available memory.

- Soundmaps take up memory, but can be kept waiting there until required. Table A.1 shows how much space one second of soundmap takes up.

Table A.1 Memory requirements for a one-second soundmap.

Level		kilobytes
A	Stereo	80
	Mono	40
B	Stereo	40
	Mono	20
C	Stereo	20
	Mono	10

Further reading

Collier D. (1991). *Collier's Rules for Desktop Design and Typography*. Wokingham: Addison-Wesley

Laurel B., ed. (1990). *The Art of Human-Computer Interface Design*. Reading, MA: Addison-Wesley

Laurel B., (1991). *Computers as Theatre*. Reading, MA: Addison-Wesley

Philips International (1987). *Compact Disc Interactive: A Designer's Overview*. Deventer: Kluwer Technical Books

Rubin T. (1988). *User Interface Design for Computer Systems*. Chichester: Ellis Horwood

Woodhead N. (1990). *Hypertext and Hypermedia: Theory and Applications*. Wilmslow: Sigma Press, Wokingham: Addison-Wesley

Glossary

Terms appearing in *italic* type within an entry have their own definitions in the Glossary.

Adaptive delta pulse code modulator All audio signals used by CD-I use a method of digital encoding known as Adaptive Delta Pulse Code Modulation (ADPCM). This involves sampling the sound and encoding the difference between successive samples. Before it can be played back the data has to be decoded.

ADPCM See *Adaptive delta pulse code modulator*

Analogue/digital Analogue and digital are two methods of recording and displaying data. When Caruso sang into the horn of a gramophone, making a vibrating needle cut directly into a wax cylinder, that was analogue recording. When Pavarotti sings into a microphone and his voice is recorded as a series of numerical values, that is digital recording. The difference is that analogue methods of recording or displaying information are continuously variable, while digital methods break the information up into small bits (pulses). A watch with hands is an analogue display, while a watch with numbers is digital. Analogue methods of recording and reproduction tend to cause degradation (for example, photocopying), while digital methods are proof against this.

Application The use of a technology (such as CD-I) for a specific purpose, often in the form of a commercial software package. For example, a 'how-to-drive' CD-I is a training application, and MediaMogul is a multimedia production application.

Aspect ratio The way in which shapes appearing on a screen are distorted by the proportions of the pixels making up the screen. Some pixels are square (for example, on Apple Macintosh machines), and some are rectangular. This means that the data that appears as a circle on one screen will appear oval on another. The mathematical expression of the distortion occurring in this way is called the aspect ratio and should be taken into consideration when designing for the screen.

Audio quality level There are four sound quality formats in CD-I. The highest is CD-DA, the quality of an ordinary CD. It uses all the available bandwidth. Level A is equivalent to the sort of sound quality you obtain when you play a brand-new LP on good-quality equipment, although without any of the background hiss. You can get two hours of stereo or four hours of mono level A sound on a CD-I disc. Level B sound is equivalent to a first-class FM radio broadcast. You can get up to four hours of stereo level B sound on a CD-I disc. Level C is more like AM radio received under good conditions. It is completely adequate for speech. You can get about nine hours of level C stereo audio on a CD-I disc.

Authoring Producing a CD-I programme, from concept to master tape. Authoring can also be used to apply more specifically to the electronic stages of the process, such as editing.

Authoring platform A combination of a machine and some software that is used to stitch together and create the links in CD-I programmes. It also allows *emulation* of the programme and may include some graphics facilities for creating *menus*, text pages and so on.

Bandwidth Bandwidth is the speed at which data can be transferred from the CD-I disc and used by the player. The CD-I player can read 75 data *sectors* per second, which is 172 kilobytes of data per second. The speed at which data is read becomes crucial when a lot of data has to be transferred – for example, when high-quality images and sound are used at the same time, or if there are several sound tracks running parallel to allow switching between them in *real time*.

Blind A two-plane visual effect in which the image on the front plane becomes like a Venetian blind that opens to reveal the image on the back plane. See *Picture planes*.

Branching Moving away from one part of a CD-I programme to another. This is usually in response to user interaction.

Byte A piece of data that is 7 or 8 bits long. 1024 bytes make one kilobyte (abbreviated as K or KB), and 1024 kilobytes make a megabyte (abbreviated as MB).

CD-DA Compact Disc Digital Audio. Uses the compact disc format for storing high-quality digital audio.

CD-I Compact Disc Interactive.

CD-ROM Compact Disc Read-Only Memory. Uses the large storage capacity of a CD disc to store digitally encoded data. This is usually text, and is used in electronic publishing for databases and reference works.

CD-ROM XA Compact Disc Read-Only Memory Extended Architecture. Uses the large storage capacity of a CD disc to store digitally encoded data. The extended architecture is used to interleave blocks of text, images, music and program data.

CD-RTOS Compact Disc Real-Time Operating System, the CD-I operating system. CD-RTOS is a flexible modularized system that can run a large number of programs at the same time, and so is particularly suited to CD-I applications.

Channel A channel is the path along which information flows from the CD-I disc to the player. CD-I can handle up to 16 audio channels and

32 channels for other purposes (for example, video, software and so on). The size of a channel is called its *bandwidth*.

Chroma key A two-*picture plane* technique that makes a selected colour or colours on the front plane transparent so that the image on the back plane can be seen through it. So, for example, a person can be filmed against a background of a certain colour, and chroma key can be used to make this background invisible so that a different background can be used.

CLUT See *Colour Look-Up Table*.

Colour Look-Up Table (CLUT) A matrix containing the colours that can be used in an image. Because each colour in the table can be referred to by its position in the table instead of by specifying its absolute colour value, this is an effective way of saving storage space on the disc. There are three types of CLUT: CLUT 8, CLUT 7 and CLUT 4. The table for a CLUT 7 image gives a choice of 128 colours, and that for a CLUT 8 gives 256 colours. This choice can be expanded by using *dynamic updating* – a useful technique in animation. Each single-screen CLUT image takes up the following space in the CD-I player's memory:

PAL or compatibility	105 kilobytes
NTSC	85 kilobytes

Further compression of CLUT images can be achieved by using *RLE*.

Compatibility mode This is the screen specification for a CD-I programme that has to run on both *PAL* and *NTSC* screens. The screen area is 384 x 280 pixels, and the *safe area* is 320 × 210 pixels.

Compression A technique for reducing the amount of data needed on the disc to store images or sounds, or sequences of them. For example, if an image has large areas that are the same colour, instead of the colour of every cell of the picture being recorded, the colour is recorded once, along with its location and the number of cells over which it extends. When the disc is played the image is decompressed before it appears on the screen. There are several different methods of compressing images, which suit different types of image, such as natural images and animated (cartoon) images. These include *DYUV*, *CLUT* and *RLE*.

Curtain A two-plane visual effect in which the image on the front plane parts or closes like a pair of curtains to reveal the image on the back plane. See *Picture planes*.

Cut A one- or two-plane visual effect in which one image disappears and is immediately replaced by a different one, either on the same plane or another. See *Picture planes*.

DAT Digital Audio Tape. A digital audio tape contains up to two hours of continuous digital sound.

DCC Digital Compact Cassette tape. Digitally recorded tape that has the same high quality as DAT, but has the advantage of using the existing standard tape cassette.

Delta Luminance Colour Difference See *DYUV*.

Digital See *Analogue/digital.*

Dynamic updating A technique used with CLUT images to increase the number of colours available. As each line of the screen is painted, the Colour Look-Up Table is changed. Thus instead of the 256 colours normally available from a single CLUT 8 image, a range of over 1000 colours can be obtained in a single image.

DYUV Delta Luminance Colour Difference, a compression technique for storing image data. Instead of recording the absolute colour value of each cell or pixel of the image, it stores only the relative differences in brightness (known as Y) and colour (known as U and V) between adjacent cells. You can have DYUV images on both *picture planes* at the same time to achieve plane effects. Each single whole-screen image takes up the following space in the CD-I player's memory:

PAL or compatibility mode	105 kilobytes
NTSC	85 kilobytes

DYUV images are suitable for storing 'natural' images, such as photographs.

Emulation The playback process used during development of CD-I titles. The title is developed on the hard disk of the computer. To play it back as if it were a real CD-I disc being played on a CD-I player, a program called an emulator is used, which enables the computer to imitate a CD-I player. This is one of the best ways of judging whether the programme is working in the way you intended.

Fade up/down A one-plane visual effect in which the image fades into blankness before being replaced by the next image. See also *Picture planes.*

Flowchart A technique used by computer software designers and others to represent sequences of events in a program, using drawings of boxes representing program elements, connected by directional lines.

Font A set of letter shapes used for displaying text – for example, Times Roman. There are two ways of displaying text in CD-I: either using the resident font of the system, which is satisfactory for utilitarian functions such as displaying system messages; or by downloading a chosen font at runtime, and using that to display text. The second technique is obviously more suitable for text displays where the appearance of the letters is important.

Full motion A video image that moves naturally and without jerkiness. Smooth motion normally requires a frame speed of at least 25 frames per second. A CD-I disc will hold approximately 72 minutes of full-motion video. This can be increased if only a part of the screen is used to display the moving image (*partial-screen* video). See also *Partial motion.*

Green Book Two volumes in which all the specifications of the standard for CD-I players and discs are contained. The standard is subscribed to by Philips, Sony and other key manufacturers, and is to CD-I what the Red Book standard was to CD-DA discs.

Hotspot An area of the screen that is used to make selections and choices in a CD-I programme. Typically, the user uses the remote control to move the pointer on the screen to the hotspot, and clicks a button to make a choice. The hotspot is often a menu button, though it can be some other object, such as an icon.

HyperCard An information tool developed by Apple for its Macintosh range of personal computers, which allows you to store different types of data (still and moving video, graphics, text, music, animation, speech) on 'cards' . The cards are held in stacks and can be accessed in a number of different ways, according to different topics and methods of classification. HyperCard can be a useful *prototyping* tool for testing the logic of a CD-I program.

Hypermedia A way of storing information so that it can be referenced and used in a nonlinear manner, one point of information being accessed directly from another without the need to go to an intervening index or table of contents. Apple's HyperCard program is probably the best-known commercial hypermedia tool.

Icon An icon is a small picture or image that stands for something. For example, an hourglass on the screen might mean that a process is going on and you have to wait. Icons are a useful nonverbal way of showing what is happening or what you need to do. You can use icons with hotspots to enliven the process of making interactive choices for the audience.

Interactive multimedia The combination of *interactivity* with *multimedia* enables the audience or user to make choices and control the pace, direction and content of a programme. CD-I is interactive multimedia.

Interactivity The flow of input and output between two systems: in the case of CD-I, between the user and the CD-I player and disc. The user's choices control the pace, direction, content and other aspects of a multimedia programme.

Interface The interface is the place where a system meets its user. This happens on two levels: first at the hardware level, where the interface is the type of equipment used (for example: remote control, keyboard, touch–screen); second at the software level, where it is the way the system appears to the user (for example: menus, hotspots and so on). A well-designed interface is essential to make a programme usable and attractive to viewers.

Interleaving A way in which the *sectors* on a CD-I disc can be arranged that allows all the information necessary for the pictures and sounds in the programme to be read at the right time. So, for example, if a programme demands parallel sound tracks in different languages, sectors making up the image and the different sound tracks will be *interleaved*, or woven together on the disc.

Kilobyte (K) A kilobyte is 1024 *bytes*. 1024 kilobytes make one *megabyte*.

Lipsynch The *synchronization* of lip movements on screen with speech sounds on the soundtrack.

Mastering The production of the master disc, from which copies can be made. See *Replication*.

Matte A two-plane visual effect in which a defined area of the front plane is made transparent, revealing the image on the back plane. See *Picture planes*.

Megabyte Approximately a million *bytes* of data. The resident memory of the CD-I player has a capacity of one megabyte, divided into 512 kilobytes for each of the two *Picture planes*.

Menu A menu is a list of items on screen from which the user can choose. In an interactive system this can be one way of enabling the user to interact. Each item is accompanied by a button; the user moves the pointer to the button and clicks the remote control to make a choice.

Mosaic Mosaics are single-*picture plane* effects that are used in images to achieve two types of result: progressive coarsening of the image until it is dissolved (pixel hold); and enlargement of the image (pixel repeat)

Multimedia A medium employing the combination and recombination of elements in a number of individual media, such as video, audio, text and graphics. The process relies upon a computer to bring these elements together and make them into a coherent product. See also *Interactive multimedia*.

National Television Standards Committee See *NTSC*.

Natural image Pictures that are photographic and appear realistic.

Non-Volatile Random Access Memory See *NVRAM*.

NTSC The standard for television pictures in the United States. (The acronym stands for National Television Standards Committee.) The picture size for images to play on NTSC television sets is 360 × 240 pixels, and the *safe area* is 320 × 210 pixels. See also *PAL* and *Compatibility mode*.

NVRAM Non-Volatile Random Access Memory. NVRAM is a small permanent memory in the CD-I player that can be used to store data that you do not want to lose when the machine is switched off. This could be, for example, password information or game scores. The CD-I player's NVRAM holds eight kilobytes of data. See also *Personal memory card*.

PAL Phase Alternating Line. The standard for television pictures in the UK, most of Europe, Australia and South America. The picture size for images to play on PAL television sets is 384 × 280 pixels, and the *safe area* is 320 × 250 pixels. See also *NTSC* and *Compatibility mode*.

Partial motion The use of a series of still images at a speed of less than approximately 20 frames per second to achieve a jerky sort of motion. This makes fewer demands upon the capacities of the CD-I player and disc, and is actually more suitable than *full motion* for some applications, such as slowed-down procedures (mending a tap) or surrogate walks.

Partial-screen video The use of only part of the screen for some purpose. This is a good way of saving disc space when you are using full-motion video. For example, if you have a window of full-motion video that takes up 40% of the screen area, you will need only 40% as much memory to store the moving video image.

Partial updating The CD-I player screen can be divided horizontally into a number of *subscreens*. Each of these can hold pictures of a different type (for example, CLUT and DYUV images can be in subscreens of the same screen), and can be updated or changed separately from the other subscreens.

Personal memory card An optional extra permanent memory for some CD-I players, which slots into the front. It can be used when an application or programme needs to store more information than the system's resident permanent memory (*NVRAM*) can hold.

Phase alternating line See *PAL*.

Picture planes Manipulation of moving and still images on the CD-I screen is enhanced by the use of the picture planes, arranged one behind another. Altogether there are four picture planes: at the very front is the cursor plane, which holds a cursor of up to 16×16 pixels; behind that are the two image planes, each able to hold an image of up to 512 kilobytes; and at the back is a plane that can be a single colour, or can be used to hold a video image from a source outside the CD-I player. The two middle planes, the image planes, are the most important. A number of effects such as *curtains, blinds* and *mattes,* can be used to change between the images on the two planes, or to show parts of images from both planes.

Pixel A contraction of 'picture element'. The basic cell or unit of a television or computer-screen picture. The number and size of pixels available on a screen dictate the resolution of the image. To make an image the colour and brightness of each cell must be specified. Pixels can be manipulated as the basis for *mosaic* picture effects. See also *NTSC, PAL* .

Pointer The small object on the screen that can be moved around – for example, by using the *remote control*. Typically, the pointer is used to point to an item on a menu, which is then selected by clicking a button on the pointing device.

Program A computer program controls and organizes the contents of the CD-I *programme*.

Programme For the purposes of this book a CD-I production, like its relatives on radio or television, is called a programme.

Prototyping As with any product, a sort of working model may the best way of checking whether it is going to work or not. When designing a CD-I disc, certain authoring platforms may be used to construct a simple prototype before production commences in earnest.

Real time Real-time areas of the CD-I programme are ones that are played as they are read. For example, audio material is read from the disc, processed and played in real time. Images first pass into the player's memory, where they are accessed as required by the control program, and are not real-time. *Interleaving* techniques for arranging data on the CD-I disc make it possible for users to switch between audio *channels* in real time. For example, if a guide-disc has parallel language tracks, it is possible to design the disc so that the user can switch between languages instantaneously.

Remote control The physical device that the CD-I viewer uses to control and interact with a programme. Typically, it looks like the remote control for a video player or a television, with the addition of a small joystick. The joystick is used to move the *pointer* around the screen.

Replication The manufacturing process by which copies of the master CD-I disc are made.

RGB Red Green Blue. The complete name for this image coding technique is RGB 5:5:5. For each *pixel* the amounts of red, green and blue are specified using five bits of data, giving 32 levels of intensity for each component, or 32,768 colours in total. This technique is suitable for very high-quality images – such as reproductions of paintings – but uses the whole of the CD-I player's memory, so that you can only use one *picture plane* with RGB images.

RLE A technique, run-length encoding, of compressing the amount of memory space needed to store the data for recording an image. This technique records only the value of the colour and the number of pixels over which it extends in each line. So a line of the screen in which there is only one colour would need only two bytes of data storage instead of one byte per pixel, as in CLUT 7 and CLUT 8 images. The degree of compression achieved by RLE depends upon the complexity of the image being stored in this way.

Rolling demo A segment of a CD-I programme that can be used at retail outlets or exhibitions to show what the programme is and what it does, and to sample its look and feel.

Run-length encoding See *RLE.*

Safe area The safe area of the CD-I screen is the area in which you can place hotspots or menu buttons where they will not be affected by the edge of the screen. The size of the safe area varies according to the television standard in use (*PAL* or *NTSC*). For CD-I programmes to play on both types of receiver, use the *compatibility mode* safe area. The safe areas are as follows:

PAL	320 × 250 pixels
NTSC and compatibility mode	320 × 210 pixels

Scrolling A useful way of presenting large images on the CD-I screen and giving the illusion of a camera panning across them. In fact what happens is that the whole image is loaded into the player's memory, and is presented piece by piece on the screen. Thus by scrolling an image of the Leaning Tower of Pisa downwards, you would present the illusion of a camera panning upwards.

SECAM The standard for television pictures used in France. The acronym stands for *Système Electronique Couleur Avec Mémoire.* As far as CD-I production is concerned it is equivalent to PAL.

Sector An area of the CD-I disc containing data. Each sector contains approximately two kilobytes of data, and the player can read 75 sectors per second. A sector can contain audio, video or text, as well as control data. Different types of sector are interleaved to assist real-time playback (see *Interleaving*). Sectors on the disc are arranged in *tracks*. The disc should be laid out so that tracks containing closely associated material are near each other, thus minimizing *seek time*.

Seek time The time taken by the CD-I player's laser pickup to move from one track or sector on the disc to another. If the disc is well laid out – that is, if tracks and sectors holding material that is closely related are arranged so that they are close to each other – seek times should be small. The maximum seek time is approximately three seconds.

Software engine A piece of software that carries out a particular task within a CD-I programme – for example, producing a dissolve between the images on the front and back picture planes or, on a higher level, arranging for a sequence of pictures with accompanying audio to be presented. The software engine is standardized and can be used repeatedly wherever required.

Soundmap A small fragment of sound that is stored in the CD-I player's memory. This can be used to reduce the amount of data the player has to read off the CD-I disc and process. The soundmap can be anything from bird-song or applause to a few bars of a tune that can be looped to make longer sound effects. Soundmaps are also useful for gongs and whistles that confirm and reinforce audience choices.

Sprite A small shape that can be moved around the screen under programme control. This can be anything from a fancy cursor or pointer to a font used for displaying text.

Square A two-plane visual effect in which the image on the front plane becomes a square opening or closing, revealing the image on the back plane. See *Picture planes*.

Storyboard The picture-by-picture 'script' for the visual aspects of a CD-I programme. Storyboarding is a technique familiar from the world of film and television.

Subscreen A horizontal division of the CD-I screen in which a specific picture type may be used. For example, different subscreens of the same screen may use *CLUT* and *DYUV* images respectively. The updating of subscreens separately and at different times is known as *partial updating*.

Synchronization For any CD-I production, sound and vision must be correctly lined up at the beginning of each sequence. So, for example, in the case of a talking head lip movements must synchronize with speech sounds (*lipsynch*).

Touch-screen A type of *interface* in which the user touches the screen of the CD-I player to make selections and control the programme. It is ideally suited to point-of-sale and point-of-information systems.

Track The *sectors* on a CD-I disc are arranged in tracks. These are a little like the tracks on an LP and are read sequentially. On a well-laid-out disc, tracks containing closely related material should be close to each other to minimize *seek times*.

Treatment The outline of a CD-I programme that is used to describe the product and to interest potential clients, publishers and so on.

Validation The process of testing a CD-I disc to make sure that it is ready to be distributed.

Video quality Video images can be produced at several levels of quality: *RGB* – suitable for images that require high quality, such as paintings or works of art; *DYUV* – suitable for natural images, such as photographs, requiring a large range of colours; *CLUT* 8 – suitable for computer-generated images, with limited colour range, and backdrops; *CLUT* 7 – suitable for animation; *RLE* – suitable for animation, graphics, text and lettering.

Wipe A two-plane visual effect in which the image on the front plane is rolled back along a horizontal line travelling vertically, or a vertical one travelling horizontally, revealing the image on the back plane. See *Picture planes*.

WORM Write-Once, Read-Many, describing a type of disc that can be recorded on only once. This type of disc is not suitable for mass production, but can be used in small editions or as trial discs for testing and validation purposes.

Index